T0313373

Thyroid Systems Engineering
A Primer in Mathematical Modeling of the Hypothalamus-Pituitary-Thyroid Axis

RIVER PUBLISHERS SERIES IN BIOMEDICAL ENGINEERING

Series Editor

DINESH KANT KUMAR
RMIT University
Australia

Indexing: All books published in this series are submitted to the Web of Science Book Citation Index (BkCI), to CrossRef and to Google Scholar.

The "River Publishers Series in Biomedical Engineering" is a series of comprehensive academic and professional books which focus on the engineering and mathematics in medicine and biology. The series presents innovative experimental science and technological development in the biomedical field as well as clinical application of new developments.

Books published in the series include research monographs, edited volumes, handbooks and textbooks. The books provide professionals, researchers, educators, and advanced students in the field with an invaluable insight into the latest research and developments.

Topics covered in the series include, but are by no means restricted to the following:

- Biomedical engineering
- Biomedical physics and applied biophysics
- Bio-informatics
- Bio-metrics
- Bio-signals
- Medical Imaging

For a list of other books in this series, visit www.riverpublishers.com

Thyroid Systems Engineering
A Primer in Mathematical Modeling of the Hypothalamus-Pituitary-Thyroid Axis

Simon Lucas Goede

Systems Research NL and TU Delft
The Netherlands

Melvin Khee-Shing Leow

Tan Tock Seng Hospital
Yong Loo Lin School of Medicine at the National University of Singapore
Duke-NUS Medical School, Singapore
Agency for Science, Technology and Research (A*STAR), Singapore
Nanyang Technological University, Singapore

River Publishers

Routledge
Taylor & Francis Group

LONDON AND NEW YORK

Published 2018 by River Publishers
River Publishers
Alsbjergvej 10, 9260 Gistrup, Denmark
www.riverpublishers.com

Distributed exclusively by Routledge
4 Park Square, Milton Park, Abingdon, Oxon OX14 4RN
605 Third Avenue, New York, NY 10017, USA

Thyroid Systems Engineering A Primer in Mathematical Modeling of the Hypothalamus-Pituitary-Thyroid Axis / by Simon Lucas Goede, Melvin Khee-Shing Leow.

Routledge is an imprint of the Taylor & Francis Group, an informa business

ISBN 978-87-93609-59-4 (print)

While every effort is made to provide dependable information, the publisher, authors, and editors cannot be held responsible for any errors or omissions.

DEDICATION

To my wife for her lasting tenacity to find the best knowledge and people to get our quest for the HPT theory in the place it currently has; our children Froukje and Janneke for their moral support and understanding; to our grandchildren Mare, Robin, Isa and Linde for their unconditional love and last but not least to Professor Dr. Jan W.A. Smit of the Radboud Medical Centre in Nijmegen, the Netherlands, for his continuous encouragement and support.

–Simon Lucas Goede

I dedicate this book to my wonderful wife, Jane Sim-Joo Tan for her unfailing love, kindness, wise counsel, emotional support and companionship; to my beloved children Rachel May-Wern Leow, Veronica May-Gwen Leow, Abigail May-Shan Leow and Eunice May-Jane Leow for their constant humor, critique, inspiration and good cheer; to my father, Alvin Yong-Keen Leow, my mother, Molly Nor-Kee Ng, my sister, Ginny Pui-Shen Leow and my uncle Lawrence Yong-Kwong Leow for their encouragement, patience, prayers and company.

–Melvin Khee-Shing Leow

*"All truth passes through three stages.
First, it is ridiculed. Second, it is violently opposed.
Third, it is accepted as being self-evident."*

–Arthur Schopenhauer, German philosopher (1788–1860)

Contents

Preface

"If mathematical equations can't make progress in understanding complex phenomena in the natural world, how might we make progress?"

–Stephen Wolfram (1959-present)

Modeling Considerations in Biological Systems

This book is a prelude to an exciting interdisciplinary field of science and medicine, namely *Thyroid Systems Engineering* where physicists, electrical engineers, chemists, biologists, endocrinologists and other related scientists work together.

Mathematical formulation in science has traditionally dominated the fields of physics and chemistry hundreds of years before our current era. Mathematical modeling of biological processes lags behind largely because the living organism is profoundly more complex with multiple variables, parameters and interactions layered within and between organ systems. Notwithstanding this, the understanding of the physiological structure and the physical modes of operation are prerequisites for any progress in fundamental research.

Today, big data research is en vogue with the hope to find a theory about the subject of research that is believed to be found with inferential statistical methods.

However, we have to realize that sophisticated modeling differs from such statistical approaches. Statistical data investigation and analysis is only sensible when it refers to identical objects in a closed system of which the properties have been formulated upfront by a theoretical framework as we encounter in statistical physics, thermodynamics and quantum mechanics.

Therefore, data alone cannot unravel behavioral mechanisms as the only clue we could find in measured data and then hope to find a recognizable relationship between variables and phenomena we have encountered. On the contrary, well known examples in physics and technology allow us to apply the physical fundamentals to find behavioral similarities that could lead to a proper first order model of the underlying physiology. When a model of human physiology has been identified, it is also necessary to realize that it is a partial description that applies to the subject or individual of investigation which prohibits to generalize any model as a simple formula valid for all people. Our individuality and uniqueness is a direct inherent consequence of our genetic makeup. This genetic individuality urges us to develop parameterized models with the minimum amount of free parameters and validate these models to the measured data of an individual. Such strategies of mathematical modeling are necessary to bridge the understanding of systems biology with physiological research.

Imagine a biological system defined as the constellation of organic, living entities of which the manifestations, as we encounter them, are the result of billions of years of evolutionary development. As such we are confronted with a complexity that seems in many cases difficult or impossible to analyze, let alone to find a suitable behavioral model from such a complexity analysis. In the analysis and synthesis of these complex functional entities, the issue of observability, testability and controllability is often overlooked or ignored. In order to analyze and synthesize complex system functions, we have to partition the complete function in minimum parts (i.e. atomic, functional and orthogonal units) that are idealistically independent of the surrounding environment. This opens the way to analyze, synthesize, design and simulate this functionality with a minimum of components. Such an atomic unit has to be designed for structural testability and controllability in order to keep track of the resulting chain of events and interactions.

A similar approach can be followed in the understanding and modeling of biological systems. The behavioral model analysis is based on stimulus and response according to systems theoretical considerations and has to be verifiable with measured data from the real biological entity. In this way, the model can be verified for correctness on primary dominant functionality. When more parts of the biological system have been modeled and verified, we have to model the interactional layers that represent the communicational interface between the modeled parts. This will be possible by the introduction of a biological network theory. Barring exceptions, it turns out that much of Nature as we encounter it in biological systems seldom manifests itself with discontinuities, such as sudden bends or similar nonlinearities. This implies that it is reasonable to treat all relational functions, dynamics etc, as continuous functions differentiable over the complete range of operation. Therefore all mathematical considerations and analysis can be based on analytical functions. Nonlinearities are here only expressed as the term indicates and refers to all nonlinear functions that are complying to the demands of analytical functions, excluding discontinuities. As we will see, most simple time-dependent biological phenomena can be characterized as the electrical network equivalent of a capacitor and a resistor. Also the important effect of amplification or gain can be encountered.

Integral to complex biological systems is the requirement of homeostasis in which thyroid physiology is a good example. Closed loop feedback systems cannot be analyzed without a fundamental knowledge of systems and control theory. A very attractive element of this method is the use of transformed differential equations which, in most cases, can be described with electrical network models based on the application of linear transforms like Laplace, Z-transform or Hilbert transform. The analysis and synthesis methods of this work are based on the knowledge of solid state physics, electrical network theory, advanced calculus, function theory, vector algebra, communication theory, system theory and signal theory.

A very important question arises when one would ask: What is the purpose of these models, and what could be the benefits? With a good model of a certain mechanism or process in the human body, we can learn to improve our understanding of the physiology and we can predict certain behavior and biological responses when we take the proper conditions into account, or change some conditions without physical invasion and burden to the human subject under investigation. Mathematical models can also address hypothetical questions that cannot be answered by conducting experiments due to ethical restrictions. Such model knowledge can provide deep insights and understanding that aid diagnosis and possibly illuminate improved or innovative ways to treat a complicated disorder. The main thyroid research issues are answered in this volume, such as the individualized relationship between homeostatic free thyroxine and thyroid stimulating hormone concentrations, the position of the homeostatic euthyroid set point and the individualized thyrotropin reference ranges.

Therefore, this book is targeted at anyone interested to understand how systems theory and mathematical modeling can be applied to endocrine physiology as illustrated by the human hypothalamus-pituitary-thyroid (HPT) system. A solid theoretical understanding forms the fundament of all possible methods for a proper diagnostic process that leads to optimum choices for the treatment of persons with a HPT malfunction.

Simon L. Goede & Melvin K. S. Leow
13th May 2017

DISCLAIMER

The authors of this book will not accept responsibility for possible injury or death etc. by following any of our recommendations in this book. Before adopting and application of treatment methods, readers should consult their own competent medical experts such as their general physicians or endocrinologists or thyroidologists specialists before adopting any of the suggestions or drawing inferences from the material presented in this book. Care has been taken to confirm the accuracy of the information presented and every effort has been made to rectify any error in this book. The authors and publisher are not responsible for oversights or omissions or for any consequences from application of any of the information in this book and make no warranty, expressed or implied with respect to the contents of the publication.

Acknowledgements

The authors would like to acknowledge the insightful inputs and skillful editorial help of Kiek Zwolsman and Jane Sim-Joo Tan. We also thank Veronica May-Gwen Leow for shepherding the illustration of the book cover into the finished version that went into production.

In the academic world this work received appreciation and encouragement from Professor Dr. Jan W.A. Smit, endocrinologist and specialist in thyroid disorders at the Radboud Medical centre in Nijmegen, The Netherlands, from discussions with Professor Dr. Wouter Serdijn and Dr. Chris Verhoeven from the University of Technology in Delft.

In addition, we would like to express our sincere appreciation to the publisher without whom this manuscript would not have evolved into the published book you now hold in your hands. We also thank the numerous thyroid patients, doctors, colleagues and students for the stimulus they provided to initiate the spark of imagination for us to believe that such a book ought to be written in the first place.

List of Figures

List of Tables

List of Abbreviations

HPT	Hypothalamus Pituitary Thyroid
TSH	Thyroid Stimulating Hormone
T3	Triiodothyronine
T4	Thyroxine
FT3	Free Triiodothyronine
FT4	Free Thyroxine
rT3	reversed Triiodothyronine
ID1	Deiodinase type 1
ID2	Deiodinase type 2
ID3	Deiodinase type 3
TRH	Thyrotropin Releasing Hormone
TSH-R	TSH receptor
TBG	Thyroid hormone Binding Globulin
TFT	Thyroid Function Test
L-T4	Levothyroxine or synthetic T4
L-T3	Liothyronine or synthetic T3
G_L	Loop gain
PK	Pharmacokinetic
PD	Pharmacodynamic

1

General Introduction

"Acknowledging what you know and recognizing what you don't know, that is knowledge."

–Lao Tzu (604–531 BC)

1.1 Introduction

In science and technology we see a myriad of different disciplines, each of which has specific research methodologies and has developed over time specific terminologies in the specific realms of its trades. As one of the oldest disciplines, medicine has followed a similar course, having evolved over many centuries and riddled with an exponentially increasing database of collective knowledge, experience, and research sophistication. Medicine deals with arguably the most precious possession we have as human beings – our health along with its associated quality of life. These matters are for all of us very important and we deal with them every day.

In the beginning of the 19th century, the situation in medicine began to change with a greater role of physics, chemistry, mathematics, and engineering within its domains resulting in the establishment of hybrid fields such as biochemistry, pharmacology, biophysics, and molecular biology. Parallel advancements in non-medical disciplines have been largely responsible for the introduction of more specialized scientific techniques and instrumentation which catalyzed quantum leaps in prevention, diagnosis, prognosis, therapeutics, and medical research. A beautiful example is the emerging technology of ultrafine mechanical gadgetry applied by a watch maker being borrowed and implemented as an application for medical diagnostics. Today, the modern medical research laboratory is equipped with contraptions like flow cytometry, immunoassays, mass spectrometers, high-performance liquid chromatography, next-generation genome sequencers, laser scanning confocal microscopy, and other "omics" technologies.

By this stage of evolution in medical science, a new cadre of scientifically and technologically educated specialists who had mastered the language of mathematics and bioengineering became pivotal in the crafting, invention, and design of current and future cutting-edge apparatuses and technologies. In particular, those who possess the skills of mathematical modeling have a distinct advantage in rendering the mysteries behind many of the hitherto unexplained phenomena and observations in the world of medicine and biology more comprehensible. This revolution is taking shape as two kinds of experts start to pioneer and drive the discipline of modeling in medicine – mathematicians or engineers with good biological insights, on the one hand, and physicians or biologists with a fair degree of mathematical maturity, on the other. Today, we have culminated in a point where the mathematically savvy can push the frontiers of biomedical research beyond what is otherwise possible. This holds true even in the present age of bioinformatics because the large number of data generated from all kinds of hi-tech measurements on the patient's body and physiology make little sense in the absence of more hypothetically driven mathematical models.

Closer to the bedside, diagnostic blood tests revealing an array of anomalies can inform the physician about the direction in which the patient should be treated. However, these tests are largely performed by laboratory analytical chemists with state-of-the-art analyzers but who understand the limitations and pitfalls of their machines and techniques. To serve the patients well, there should be a nice cross-talk and cooperation between the clinical analytical chemist/chemical pathologist and the well-informed physician/biological researcher so that the correct diagnostic interpretation of the results can be achieved. These tests are useful only when the diagnostician knows what to expect and how to proceed when a certain result occurs. The latter is possible only when the underlying pathology is known and understood in order to treat the patient accordingly.

When we are zooming in on the subject of our book, the human hypothalamus–pituitary–thyroid system, we can learn about the knowledge and skills many researchers in the field of biological modeling and endocrinology have acquired to identify and treat the various thyroid disorders. Currently, most of the known thyroid-related disorders are treated with methods based on evidence from the published literature, clinical practice guidelines, and best local practices which are in most cases results based on clinical experience and the accepted normal ranges of thyroid function tests.

Depending on the situation, such a strategy can work well for one patient but disappoint another in terms of the intended outcome. Often, physicians apply laboratory test results to manage patients by pattern recognition and rule-based algorithms without deep insights into pathogenesis or mechanistic knowledge of the underlying disease process. The mindless application of biochemical and immunoassay reference ranges without due introspection is partly responsible for misdiagnosis and suboptimal treatment outcomes. In this respect, reference ranges encompassing fold-differences between the upper and lower limits should lead to a healthy skepticism.

These scenarios and examples urged us to develop a new theoretical framework on which personalized medicine and treatment will deliver the necessary results.

Such scenarios of personalized medicine will deliver the greatest benefits.

Even experienced thyroidologists who are probably better able to interpret and tackle thyroid function test results more proficiently than the general specialist of internal diseases can have benefits when he or she is able to learn about the content and meaning of the subjects in this book. Electrical engineers dedicating their knowledge and experience to biological modeling, working together with clinical experts, open the way to more effective models that could easily be applied on individual patient's data and develop a theoretical framework for practical applications and treatment in thyroid disorders.

This book serves as a fine example of *thyroid systems engineering* by integrating current medical science, biology, engineering, and mathematics with clinical practice. It is the product of a duo that shares the vision of dissecting thyroid physiology with the tools of mathematics so that many thyroid patients round the world may receive better targeted treatment with improved quality of life. It will come as no surprise if the contents of this book are not immediately found applicable at the bedside because the vast majority of clinicians tend to stay within the safety of what are best practices in the mainstream bolstered by guidelines and distilled from the essence of evidence-based medicine rather than be seen as mavericks with outlandish ideas and straying out of the box. It is our hope that the scientific establishment round the world can appreciate the value of mathematical modeling as applied to medicine so that these ideas as expounded within the pages of this book may find their way to better clinical outcomes in the near future.

2

Physiology of the HPT Axis

"If we knew what it was we were doing, it would not be called research."

−Albert Einstein (1879–1955)

2.1 Synopsis of Thyroid Biochemistry and Molecular Biology

Thyroid hormones govern a host of functions of paramount importance for survival, including general cell and tissue metabolism, regulating growth, maturation and differentiation, as well as thermogenesis. Literally, every cell and tissue of the human body finds some form of interaction with thyroid hormones due to their influence on an extensive repertoire of gene expression across a broad range of organ systems. The thyroid gland is a component of the hypothalamus–pituitary–thyroid (HPT) axis whose homeostatic control system is responsible for the maintenance of an optimal thyroid hormonal equilibrium termed the euthyroid state. This is achieved by a negative feedback loop which implies the existence of a stable set point analogous to a set temperature controlled by a feedback thermostat in an air-conditioned room. Such a euthyroid set point value is probably uniquely determined and programmed according to our individual genetic makeup [1].

In the human HPT axis, the thyroid produces mainly the hormones 3,5,3′,5′-tetraiodothyronine (T4) and 3,5,3′-triiodothyronine (T3) in the approximate ratio of 10:2. Being highly hydrophobic, these thyroid hormones must be bound to hydrophilic carrier proteins so that they can be transported in the bloodstream which forms the nexus between all critical organs so that they can in turn exert a broad range of genomic and extra-genomic effects [1].

For a clear understanding, it is helpful to examine the biosynthesis of thyroid hormones. Updated knowledge of this area indicates that apart from T4 and T3, there exist a wider range of thyroid hormone precursors and metabolites including 3,3′,5′-trioiodothyronine or reverse T3 (rT3), 3,5-diiodothyronine (3,5-T2), 3,3′-diiodothyronine (3,3′-T2), 3-iodothyronine (T1), 3-iodothyronamine (T1AM), thyronine (T0), tetraiodothyroacetic acid (Tetrac), triiodothyroacetic acid (Triac), and diiodothyroacetic acid (Diac). Although it remains doubtful if any of these possesses significant thyromimetic properties, increasing evidence suggests that some may exert effects on the thyroid hormone receptor while others may play a role in thyroid autoregulation via rapid surface membrane-mediated signaling through trace amine-associated receptors. In the interest of simplicity, we will not discuss these endogenous decarboxylated and deiodinated derivatives of T4 and T3. In Figure 2.1, the chemical presentation of T4 and the deiodination processes to T3, rT3, and T2 are depicted [2].

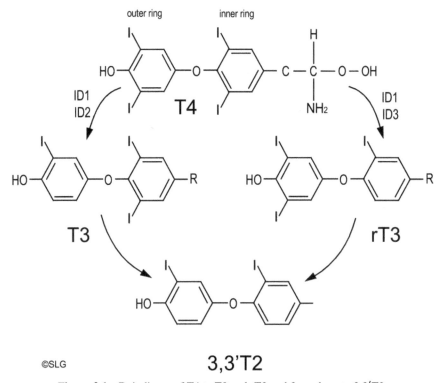

©SLG

Figure 2.1   Deiodinase of T4 to T3 and rT3 and from there to 3,3′T2.

T3 is finally produced from an amount of bioavailable T4 via 5′-deiodinase.

These processes of deiodination are locally active intra-thyroidally and control the thyroid production of T3 and rT3 in addition to other peripheral tissue deiodinases, like specific deiodinases in the heart, lungs, muscles, liver, brain, etc.

We can distinguish three types of deiodinases:

1. Type 1 or D1 commonly found in the liver and kidney can deiodinate the inner tyrosyl ring and/or the outer phenolic ring as indicated in Figure 2.1.
2. Type 2 or D2 is found in the heart, brain, skeletal muscles, adipose tissue, thyroid, and pituitary gland.
3. Type 3 or D3 in fetal tissue, placenta, and the brain, except the pituitary.

2.2 Operation of the Hypothalamus–Pituitary (HP) System

In the HPT axis, the hypothalamus operates together with the pituitary as the central control organ for the maintenance of the set point value of free T4 concentrations denoted throughout the rest of this discourse as [FT4]. The detection of FT4 at the level of the hypothalamus and pituitary is based on T3 whose concentration depends on the local relative activities of the deiodinases, D2 and D3 [3, 4]. This results in biologically significant concentrations of intracellular unbound T3, which is at least an order of magnitude higher than normal plasma [FT3] [5–8]. The unbound T3 binds to the nuclear thyroid hormone receptor TRβ2 in the hypothalamic neurons which then determines the relative expression and secretion of the neural tripeptide thyrotropin releasing hormone (TRH) or thyroliberin via the long Fekete–Lechan feedback loop [6] that is presented to the pituitary thyrotrophs after which these cells release a proportional amount of thyroid stimulating hormone (TSH) or thyrotropin. This release is also controlled by the local deiodination of unbound T4 to T3 via Dio2 (or D2) in the pituitary, which provides a secondary modulation of the TSH secretion via the classic Astwood–Hoskins feedback loop. Interestingly, an autocrine ultra-short feedback loop (Brokken–Wiersinga–Prummel loop) [9] in which TSH modulates its own secretion by binding to TSH receptors (TSH-R) on thyrotrophs itself has been proved to occur and partly responsible for the phenomenon of hysteresis observed in pathological states of thyroid hormone excess and deficiency. TSH secreted into the systemic circulation then acts on the surface TSH-R expressed on the cell membranes of the thyroid follicular cells to stimulate the synthesis of thyroid hormones. The above feedback loops constitute the key regulatory checkpoints of the HPT axis.

2.3 Thyroid Physiology

Thyroid stimulating hormone has a half-life of about 30–60 min and stimulates the thyroid gland to secrete the hormone thyroxine T4, which has only a slight effect on metabolism. The free fraction of T4, FT4, is converted into T3 intrathyroidally and peripherally. The free fraction of T3, FT3, visa deiodinase D1 or D2 is the active thyroid hormone that stimulates metabolism. About 80% of this conversion is found in the liver and other organs, and 20% in the thyroid itself. D3 is responsible for an inactivation of FT4 resulting in rT3 [10, 11].

The thyroid gland has TSH-R primarily found on the surface of the thyroid follicular epithelial cells to respond to TSH. These cells produce the thyroid hormones T4 and T3 when TSH-R is triggered. Either TSH or other thyrotropic molecules such as TSH receptor autoantibodies (TRAb) or even more recently discovered thyrostimulin (a heterodimer of glycoproteins α2β5) can bind and stimulate TSH-R. Normally, TSH is the triggering signal. However, the moment that substances other than TSH will trigger the release of thyroid hormones, the control of the HPT system can be overridden and the thyroid can then produce T4 and T3 in an uncontrolled manner. This manifestation is found in Graves' disease.

The thyroid epithelial cells take up iodine and amino acids from the blood. They synthesize thyroglobulin and thyroperoxidase from amino acids which are important in the process of thyroid hormone biosynthesis. Thyroid hormones are transported throughout the body where they control metabolism. Every cell in the body depends upon thyroid hormones for regulation of their metabolism.

2.4 Physiological Feedback Representation of the HPT System

In Figure 2.2, the physiological feedback mechanism of the HPT is presented.

The HPT system operates according to negative feedback principles. One of the properties of the HP unit is that there exists an internal reference value [12] aimed to maintain a specific set point for the FT4 concentration [13].

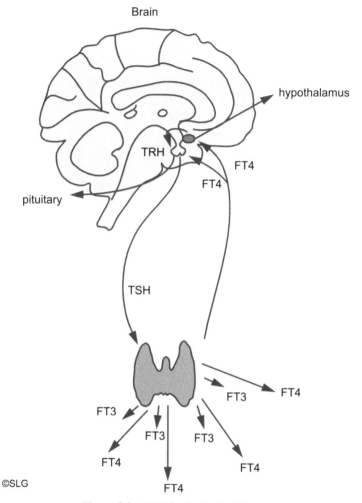

Figure 2.2 HPT feedback physiology.

This is the main task of the HPT system. As argued before, the negative feedback mechanism is based on a degree of inhibition of TSH. The higher the deiodinase activities from circulating FT4, resulting in relative high local concentrations of intracellular free T3, the stronger the inhibition of TSH. The TSH concentration is, according to this mechanism, inversely proportional to the value of the plasma FT4 concentration. [FT4] and the belonging local deiodinase processes influence the HP system which operates as a sensor with an internal reference value for [FT4] in euthyroid condition. This feedback system is directly comparable to a central heating system at home. In this sense, the temperature is represented by [FT4] and the control signal to the central heater is represented by [TSH]. When the system is in equilibrium, the production of thyroid hormones is in compliance with the demand of these substances, resulting in the maintenance of the set point value of [FT4] to the internal set reference value in the HP system. In a healthy person, this condition of equilibrium is defined as *euthyroid homeostasis* and results in the preset value of [FT4] and the belonging value for [TSH], also known as the set point. A more formal and detailed analysis will be presented in the following chapters.

References

[1] Thyroid Disease Manager (2017). Available at: http://www.thyroidmana ger.org/Physology of the hypothalamic pituitary thyroid axis.

[2] Miot, F., Dupuy, C., Dumont, J., and Rousset, B. (2015). *Chapter 2 Thyroid Hormone Synthesis and Secretion.* Available at: www. endotext.org

[3] Nillni, E. (2010). Regulation of the hypothalamic thyrotropin releasing hormone (TRH) neuron by neuronal and peripheral inputs. *Front Neuroendocrinol.* 31, 134–156. doi: 10.1016/j.yfrne.2010.01.001

[4] Saravanan, P., Siddique, H., Simmons, D. J., Greenwood, R., Dayan, C. M. (2007). Twenty-four hour hormone profiles of TSH, free T3 and free T4 in hypothyroid patients on combined T3/T4 therapy. *Exp. Clin. Endocrinol. Diabetes* 115, 261–267. doi: 10.1055/s-2007-973071

[5] Lechan, R. M., and Fekete, C. (2005). Role of thyroid hormone deiodination in the hypothalamus. *Thyroid* 15, 883–897.

[6] Reed Larsen, P., and Zavacki, A. M. (2012). Role of the iodothyronine deiodinases in the physiology and pathophysiology of thyroid hormone action. *Eur. Thyroid J.* 1, 232–242 doi: 10.1159/000343922

[7] Bianco, A. C., Brian, W., and Kim, B. W. (2006). Deiodinases: implications of the local control of thyroid hormone action. *J. Clin. Investig.* 116, 2571–2579. doi: 10.1172/JCI29812

[8] Gereben, B., Zavacki, A. M., Ribich, S., Kim, B. W., Huang, S. A., Simonides, W. S., et al. (2008). Cellular and molecular basis of deiodinase-regulated thyroid hormone signaling. *Endocr. Rev.* 29, 898–938.

[9] Prummel, M. F., Brokken, L. J. S., and Wiersinga, W. M. (2004). Ultra short-loop feedback control of thyrotropin secretion. *Thyroid* 14, 825–829. doi: 10.1089/thy.2004.14.825

[10] Werneck de Castro, J. P., Fonseca, T. L., Ueta, C. B., McAninch, E. A., Abdalla, S., Wittmann, G., et al. (2015). Differences in hypothalamic type 2 deiodinase ubiquitination explain localized sensitivity to thyroxine. *J. Clin. Invest.* 125, 769–781.

[11] Kuiper, G. G., Kester, M. H., Peeters, R. P., and Visser, T. J. (2005). Biochemical mechanisms of thyroid hormone deiodination. *Thyroid* 15, 787–798.

[12] Goede, S. L., Leow, M. K., Smit, J. W. A., Klein, HH, Dietrich JW, Hypothalamus-pituitary-thyroid feedback control: implications of mathematical modeling and consequences for thyrotropin (TSH) and free thyroxine (FT4) reference ranges. *Bull. Math. Biol.* 76, 1270–1287. doi: 10.1007/s11538-014-9955-5

[13] Leow, M. K., and Goede, S. L. (2014). The homeostatic set point of the hypothalamus-pituitary-thyroid axis – maximum curvature theory for personalized euthyroid targets. *Theor. Biol. Med. Model.* 11:3. doi: 10.1186/1742-4682-11-3

3

Modeling Principles

"Truth is stranger than fiction, but it is because fiction is obliged to stick to possibilities; Truth isn't."

–Mark Twain (1835–1910)

3.1 Introduction

In this chapter, we discuss and describe the basics of endocrine modeling. Modeling in endocrinology depends on sound understanding of developmental events, proliferation, growth, differentiation, physiology, biochemistry, and metabolism. Although the endocrine system is a complete and integrated entity, it can be compartmentalized in distinguishable sub-systems that have defined interactions with each other. Specializations include, among others, mainly endocrine sub systems like the hypothalamus–pituitary–thyroid (HPT) system, the glucose–insulin control system, hypothalamus–pituitary–gonadal (HPG) system, and the hypothalamus–pituitary–adrenal (HPA) system. Basic ideas about modeling can be found in very simple examples from day-to-day life that can also be applied on many occasions and situations in physiology. Human physiology as we know is the product of evolution over the millennia and is generally far too complex for a comprehensive modeling exercise. A very important first step is to begin by observing some aspects, relationships, and effects of this physiology.

In biological research, a wide range of different models – phenomenological, computational, and of course, mathematical – can be applied on various areas of biology. In this book, we obviously concern ourselves with mathematical models. Even then, there are different forms of mathematical modeling including linear/non-linear, discrete/continuous, deterministic (e.g., dynamic/static), and statistical (data-driven, Bayesian inference/correlations).

Well-known examples of mathematical models being applied with much success are those of predator–prey models predicting the impact on the population equilibrium, the propagation of infectious diseases in a population, and also physiologic behavioral models in living beings.

In this treatise, the key objectives are to build a model with logical descriptions of realistic physiological inter-relationships, validate with observed data, and translate it into clinical applications. If a model cannot satisfy any of these three pillars, the model is considered an unproven theory or a fantasy. Mathematical modeling is intended to be transparent and understandable. Every attempt should be made to avoid unnecessary mathematical complexity, bearing in mind an Occam's razor approach, which seeks to find a model that balances mechanistic realism and mathematical simplicity. A common pitfall is the introduction of unobservable and untestable parameters. In the literature, there are many models "tested and validated" with various parameters. An outstanding example is the general theory of relativity in which Einstein introduced a cosmological constant to produce a steady state universe based on his erroneous conviction. A few years later in 1929, Hubble demonstrated an expanding universe by measuring the speed of distant galaxies based on their spectral red shift component. Einstein himself subsequently indicated this as a major blunder. In later theoretical developments in the other chapters, we encounter a similar set of numbers, known as the HPT set point, from which the complete euthyroid HPT behavior can be derived. When a theory connects with reality, its value is significantly elevated. This is what a usable model should do!

Besides the predictive value of a model, it can also be used as a tool for simulation. There are many situations where the effect of a model can be tested in a virtual environment without the negative side effects in a real situation. In this sense, the simulation of a complicated electronic circuit before the design is realized in a very costly manufacturing is of fundamental importance, because a single failure can result in the bankruptcy of the design company.

This demonstrates the necessity, power and advantages of simulations. Similarly, the effects of certain medicine could be tested without the use of a living being. This can only be done with a correct model and a model parameter characterization of the living entity in question. These important consequences of modeling and simulation will be discussed in the separate chapters about the HPT system. We will illustrate the use of *in silico* (computer) simulations and standard electrical network simulators for further investigations.

Regarding the endocrine and pharmacokinetic aspects, the human body can, in its simplest way, be considered as a liquid container with all kinds of compartmentalized metabolic processes necessary for survival. A very important notion here is the state of equilibrium or homeostasis. This state is maintained under the condition that the amount of input is in balance with the amount of output or vice versa in such a way that the steady state will only slightly vary around a certain value of equilibrium. System instability on the other hand can result in biological disorders and even death.

This notion of homeostasis is the basis of all modeling subjects that we will discuss.

Homeostasis can be achieved either by means of compensation or when a certain form of signal gain is involved, also negative and/or positive feedback.

Depending on whether the feedback is negative or positive, a balance in the dynamics must occur so that the system operates under conditions critical to life. An important hallmark of the endocrine system in vertebrates is that many hormones controlling a host of physiological and metabolic processes that are tightly controlled via such feedback systems. From this basis, we also encounter temporary alterations when a certain disturbance occurs. The hormonal milieu of a healthy HPT system can temporarily be perturbed by externally administered T4 or T3. Pharmacokinetic analysis shows that these perturbations are evened out to normal homeostatic conditions when given time for the body to adjust. In a separate section, we will discuss these so-called dynamical processes.

3.2 Modeling Examples

3.2.1 Modeling Example 1

We observe a container filled with water as depicted in Figure 3.1.

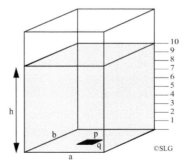

Figure 3.1 Leaking water container.

From Figure 3.1, we see that the volume of the water is represented by $V = a.b.h$ cm^3, where the area of the bottom equals $a.b$ cm^2 and h as the height of the water column. At the bottom of the container is an opening with area $p.q$ cm^2. When the opening $p.q$ is opened or unlocked, the water level will drop. The water current through the opening at the bottom is dependent on the height h of the water column. This implies that the water current will decrease as a function of time because the level h decreases as a function of time. When U represents the water pressure on the opening $p.q$ and $R = A/p.q$ cm^{-2} represents a resistance against the water pressure, we can model the process as a time-dependent voltage U across a resistor as is well known from Ohm's first law:

$$I = \frac{U}{R} \tag{3.1}$$

The current I equals the voltage U (water pressure) divided by the value of the resistor R.

We can state that the voltage U is directly linearly proportional to the value of h.

Furthermore, we can define the resistor R as being proportional to the inverse of the area $p.q$ cm^2 of the opening at the bottom. Thus:

$$R = \frac{A}{p.q} \tag{3.2}$$

The resistor is a measure of the time necessary to produce a certain amount of water or as a unit: sec/L. The inverse of the resistor is the conductance $1/R$ and expresses the amount of water we could expect after 1 s, or L/sec.

The current I equals then

$$\frac{U}{R} = h.p.q \tag{3.3}$$

As we can observe from Figure 3.1, we see that the area of the container bottom $a.b$ cm^2 and the area of the opening in the bottom $p.q$ cm^2 will remain constant. Let us then define that $a.b = C$ cm^2 and as we use (Equation 3.2); then the water level h is divided into 10 equal parts which will serve as level measurement indicator. The only things we know about this experiment are the dimensions of the water container bottom $a.b$ cm^2, the starting level of the water in the container $h,$ and the area of the hole in the bottom $p.q$ cm^2. The variables we can measure are the time in seconds and the height of the water level h. Now we will start to observe how much time is needed when we unlock the opening $p.q$ in the bottom and close it again when the water

Table 3.1 Measured data from experiment 1

Time in Seconds	H
0	10
1	9
2.2	8
3.5	7
5.1	6
6.9	5
9.1	4
12	3
16	2
22.9	1

level has dropped from 10 to 9. We will repeat this procedure until the water level has dropped to the scale indicator 1. The measured results are noted in Table 3.1.

The time was measured with a stopwatch with a reading accuracy of ±0.05 s.

From Figure 3.2, we can appreciate that the water level drop as a function of time is not according to a straight line, or linear, and that the time necessary to drop to a lower level than 1 cm will increase rapidly. In this example,

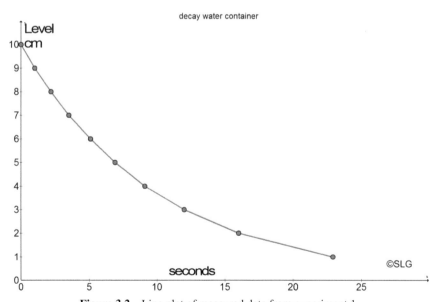

Figure 3.2 Line plot of measured data from experiment 1.

the measurement results, indicated as black dots, are here interconnected by straight lines to get a graphical impression of the relationship between time and water level when the bottom hole is opened. It would be interesting if we could find a mathematical expression for the plotted curve. If we know nothing of the properties of the container etc. in this experiment, there is the possibility to use math tools that can deliver the expression of a function that will closely fit on the measured data. With a simple free math tool downloadable from the Internet (*Graph 4.4*), we can find the answer [1].

With the measured data dots of our experiment, we will try to find a function that possibly fits. This experiment can be analyzed with several different pre-programmed parameterizable functions available in *Graph 4.4*.

For example, if we would suppose the function to be linear (which is obviously not the case here), we could propose a form like $y = ax + b$. The straight line, or linear function, has two parameters a and b.

Let us see what a fitting action of *Graph 4.4* will provide with the form

$$y = ax + b \tag{3.4}$$

From Figure 3.3, we can read that the linear model fits more or less and can be considered as a linear regression result of the measurements. The tool *Graph 4.4* gives $y = -0.3948x + 8.6074$ as the solution for the

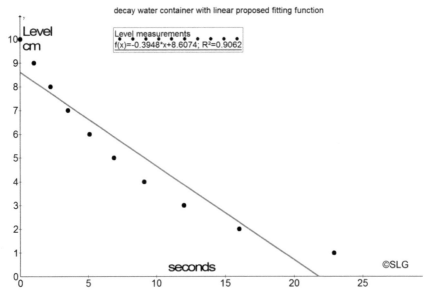

decay water container with linear proposed fitting function

Level measurements
$f(x) = -0.3948 \cdot x + 8.6074; \ R^2 = 0.9062$

©SLG

Figure 3.3 Fitting the measured data with a supposed linear function.

relationship between time and water level drop. The fitting quality is indicated as $R^2 = 0.9062$, which means that the fitting result is about 90% which is not bad for a first fitting attempt.

A second trial is performed with a power function in the general parameterized form:

$y = A.x^b$ the results of which are plotted in Figure 3.4.

Here, the *Graph 4.4* tool results in $A = 13.5057$ and $b = -0.656$.

The fitting quality is only $R^2 = 0.5807$, which is about 58%. It is obvious that the power function is not very close to what we measured.

A third trial is performed with a polynomial function in the form:

$$y = ax^2 + bx + c \qquad (3.5)$$

This function is depicted in Figure 3.5.

Because this form is polynomial, the number of coefficients is equal to the order of the polynomial plus one. In this case, these would be a, b, and c.

Obviously, the fitting results are close enough to apply a polynomial of the second order. *Graph 4.4* gives the following values: $a = 0.0182$, $b = -0.7907$, and $c = 0.0182$. The fitting quality is $R^2 = 0.9966$, which is more than 99%, actually a very good fit. This approximation gives us a rather solid idea of the

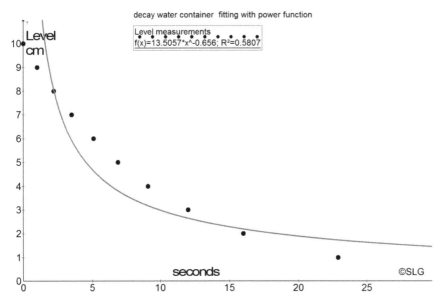

Figure 3.4 Relationship between time and water level drop with a parameterized power function $y = A.x^b$.

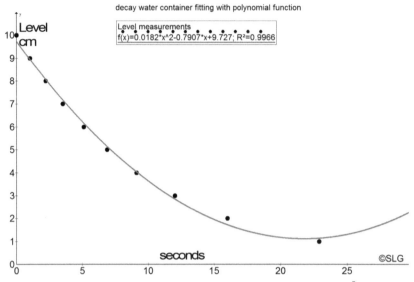

Figure 3.5   Fitting results with the use of a polynomial function $y = ax^2 + bx + c$.

behavior over the measured interval. The deviations seem to increase from 23 s and higher.

A fourth trial is performed with an exponential function in the form:

$$y = A \, \exp{(bx)} \tag{3.6}$$

The results are plotted in Figure 3.6.

From Figure 3.6, we can appreciate that the fit is perfect, $R^2 = 1$, a 100% match.

The function $y = (9.9815).(0.9045)^x$ can be re-written as $y = 9.9815 \, \exp(-0.1003x)$.

We find the exponential factor by $\ln(0.9045) = -0.1003$. So we found the model parameters and. $A = 9.9815$ and $b = -0.1003$.

With this result, we can actually do a lot more. The description of experiment 1 has all the ingredients to derive a formal model in the form of a time-dependent differential equation.

Because the leaking water container behaves like a charged electrical capacitor connected with a discharging resistor, we can use the modeling methods common in electrical network theory. From this differential equation, we can find the parameters that represent the resistor and the capacitor from which we can derive the actual dimensions of a and b for the container bottom area and the dimensions of the hole by p and q.

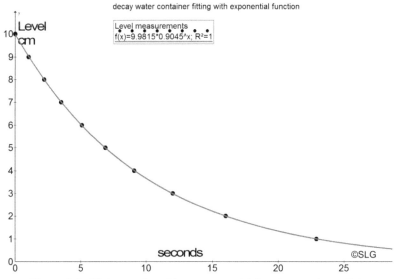

Figure 3.6 Fitting results with an exponential function $y = A \exp(bx)$.

The simplest way to solve this problem is the use of a model analogous to the function of an electrical network with a capacitor and a resistor. The reason for the electrical network is the fact that the number of related physiological and functional behavioral elements can easily be expanded with other network elements and the analysis can be transformed to the complex frequency domain and after analysis transformed back to the time domain. This method of analysis will be discussed in a separate section. Besides the possibilities to analyze higher physiological complexities with electrical network modeling, we also find the direct translation to known physiology.

3.3 Electrical Network Representation of the Leaking Water Container

From Figure 3.1, we could derive the capacitor value

$$C = a.b \tag{3.7}$$

and the resistor

$$R = \frac{A}{p.q} \tag{3.8}$$

The schematic representation is given in Figure 3.7.

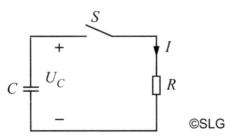

Figure 3.7   Network representation of the water container.

We use Ohm's law to analyze the static circuit

$$U = I.R \tag{3.9}$$

or

$$I = \frac{U}{R} \tag{3.10}$$

where U represents the voltage across the resistor R and I represents the value of the current in the circuit.

From our original configuration in Figure 3.1, the voltage is represented by the value of h of the water level. The action to open the opening $p.q$ in the bottom is represented by the closing of the switch in Figure 3.7. The current I is defined as the flow rate of the water as a function of time.

The total starting volume of the water container is

$$V = h.a.b \tag{3.11}$$

This is equivalent to the notion of total amount of charge Q in capacitor C. The voltage on the capacitor is equivalent with the water level h. This results in the following equation:

$$Q = C.U_c \tag{3.12}$$

The total amount of charge in the capacitor (water volume) equals the capacitor value times the height of the ware level. The average water volume variation or average charge variation as a function of time can be written as:

$$\frac{Q_1 - Q_2}{t_1 - t_2} = \frac{\Delta Q}{\Delta t} = I \tag{3.13}$$

or rewritten for infinitesimal small differences as

$$\lim_{\Delta \to 0} \left(\frac{\Delta Q}{\Delta t} \right) = \frac{dQ}{dt} \tag{3.14}$$

This implies that we differentiate the function of $Q(t)$ with respect to the variable of the time t.

When we apply that to (3.13), we find:

$$I = \frac{dQ}{dt} = -C\frac{dU_c}{dt} \tag{3.15}$$

The minus sign in the equation comes from the fact that the capacitor is discharged.

Equation (3.15) represents the relationship between capacitor current I and belonging capacitor voltage U_c.

From Figure 3.7, we read the value of I when the switch is closed:

$$I = \frac{U_c}{R} \tag{3.16}$$

When we combine Equations (3.15) and (3.16), we find

$$\frac{U_c}{R} = C\frac{-dU_c}{dt} \tag{3.17}$$

resulting in

$$U_c = -RC\frac{dU_c}{dt} \tag{3.18}$$

or

$$\frac{dU_c}{U_c} = \frac{-1}{RC}dt \tag{3.19}$$

Equation (3.19) is a first-order linear differential equation that can be solved by integration of both sides.

Then we find,

$$\int \frac{dU_c}{U_c} = \int \frac{-1}{RC}dt \tag{3.20}$$

resulting in

$$\ln(U_c) = \frac{-t}{RC} + K \tag{3.21}$$

Because the product RC has the dimension of time, we can define the time constant:

$$\tau = RC \tag{3.22}$$

Equation (3.21) can then be written as

$$U_c = \exp\left(\frac{-t}{\tau} + K\right) = \exp(K).\exp\left(\frac{-t}{\tau}\right) \tag{3.23}$$

From the boundary conditions, we can calculate the value of

$$\exp(K) = U \tag{3.24}$$

For the time-dependent form of the capacitor voltage $U_c(t)$, we find

$$U_c(t) = U \exp\left(\frac{-t}{RC}\right) \tag{3.25}$$

We can find that $1/RC = 0.1\ \text{s}^{-1}$ according to the characterized model results from experiment 1.

This can be translated into the belonging values that were derived from the container bottom dimensions and the dimensions of the hole in the bottom.

We found

$$R = \frac{1}{p.q}\ \text{seconds/cm}^2 \tag{3.26}$$

and

$$a.b = C\ \text{cm}^2 \tag{3.27}$$

Then,

$$RC = \frac{a.b}{p.q}(\text{s}) \tag{3.28}$$

or in this case

$$\frac{a.b}{p.q} = 10(\text{s}) \tag{3.29}$$

When the dimensions of a and b are known, the belonging area represented by $p.q$ can be calculated and vice versa.

3.4 Static Modeling

In the previous part, we discussed the principles of dynamic modeling. However, dynamic models are bound to time-dependent phenomena. Other models can be described as static or functions of variables. In these models, time plays no role whatsoever. It is often observed that it seems problematic to distinguish time-dependent phenomena from functional or static phenomena. We have to realize that one of the most important scientific developed models was related to the motion and position of planets on our solar system. The planetary dynamics were of course described as time functions.

In biological systems, where many phenomena have a dynamic character, there are also many that display a so-called static, parametric relationship. When we encounter homeostatic conditions of a system, the system is in equilibrium, although this equilibrium is maintained by dynamical mechanisms like steady-state currents maintaining a stable level of a leaking container. The previous container model was in that sense a good example.

The following model example describes the relationship between voltage across a resistor and the resulting electrical current through the resistor. Here, the electrical current represents the movement and flux of a number of electrical charges, namely electrons.

The electrons (in the steady state of the electrical current) represent a constant number of charges passing a defined area per second. In the electro dynamical sense, we can write

$$I = \frac{dQ}{dt} \tag{3.30}$$

where Q represents the summation of all charges passing a defined flat area. We have to realize that this notion of steady state dynamics plays a dominant role in biological systems. With this notion, we can avoid unnecessary differential equations to find simple and elegant relationships as we will discuss later.

The electrical network of Figure 3.8 is the basis for the following model description.

The current through the resistor R is defined as:

$$I = \frac{U}{R} \tag{3.31}$$

The relationship between I and U is now defined by the parameter $1/R$.

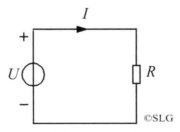

Figure 3.8   Resistor network.

In Chapter 6, we will discuss the more complex relationship between [FT4] and [TSH]. This is also a static characteristic consisting of emerging dynamic equilibriums between [FT4] and [TSH].

3.5 Modeling Examples from Solid-state Physics and Electronic Engineering

In semiconductor technology, the physical modeling about the conductivity properties in P and N doted semiconductor materials is determined by the underlying quantum mechanical theory of solid-state physics [3].

If we want to know all the conductivity mechanisms and dynamics on atomic level, we will not find a practical solution to work with on component level.

The actual response of the semiconductor component is the integrated result of all underlying atomic effects, similar to what we encounter in thermodynamics where the effect of an individual atom or molecule cannot be taken into account.

In electronic engineering, we examine the transfer and conductivity behavior on a macro level where the state of conductivity is examined after an emerging equilibrium situation. Processes in semiconductors are characterized by ultrafast transients working on a nanosecond scale. Every time the current through a PN junction in conductive mode is changed, we observe, after a very short time, the new equilibrium situation resulting in a change of voltage over the PN junction.

The conductivity properties of the PN junction are then modeled in the sense of a collection of all possible equilibrium situations related to the various current values through that junction. The elementary model expression can be derived from solid-state physics theory. A parameterized model can be used to characterize the actual conductivity behavior.

The resulting expression, a validated model for the collection of equilibriums in a junction consisting of areas of P and N conducting material, is derived from conducting and concentration properties of the belonging junction [3].

$$I_{PN} = I_s \left\{ \exp\left(\frac{qU}{kT}\right) - 1 \right\} \tag{3.32}$$

$$\frac{I_{PN}}{I_s} = \left\{ \exp\left(\frac{qU}{kT}\right) - 1 \right\} \tag{3.33}$$

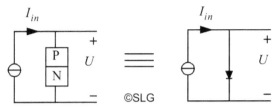

Figure 3.9 Electrical network representation of a conducting PN junction.

because $\frac{I_{PN}}{I_s} \gg 1$ we can write

$$\frac{I_{PN}}{I_s} = \exp\left(\frac{qU}{kT}\right) \tag{3.34}$$

then

$$\frac{qU}{kT} = \ln\left(\frac{I_{PN}}{I_s}\right) \tag{3.35}$$

and

$$U_{PN} = \frac{kT}{q} \ln\left(\frac{I_{PN}}{I_s}\right) = 0.26\ln\left(\frac{I_{PN}}{I_s}\right) \tag{3.36}$$

In Figure 3.9 the P/N junction together with the network element is depicted

I_{PN} represents the forward current through the PN junction and is dependent on

U_{PN} represents the voltage across the PN junction

q represents the elementary charge of an electron. $q = 1.6021 \times 10^{-19}$

k represents Boltzmann's constant $k = 1.3805 \times 10^{-23}$ J/K

T represents the absolute temperature (300°K)

The parameterization of this model is defined by the model parameter A_0, which in itself is again a function of the absolute temperature.

$$I_s = A_0 \exp[k(T_j - T_0)] \tag{3.37}$$

A_0 is a constant to be determined at a junction temperature of 300°K

T_j = actual junction temperature

$T_0 = 300°K$ (25°C)

The PN characteristic can be depicted as presented in Figure 3.10.

The previous analysis presented the static relationships between voltage and current of an exponential model of the bipolar PN diode.

When voltages and currents are applied as signals with a variation in value as a function of time, we have to apply dynamic analysis methods, which will be discussed in the following chapters.

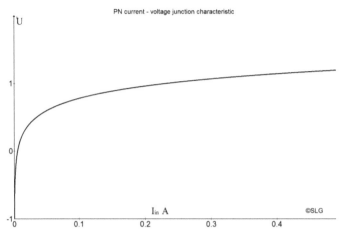

Figure 3.10   PN junction $I - U$ characteristic at a junction temperature of $300°$K.

3.6 General Appearance and Decay Model with Electrical Network Elements

In Figure 3.11 a generalized physiologic appearance and decay model is depicted.

The capacitor C represents the summation of all compartimentalized accumulation capacity, in liters, as unit of volume in the blood serum to accumulate and is supposed to be characteristic of every individual and dependent on weight.

The concentration level has the equivalent dimension of voltage.

Via the resistor R_1, with the dimension of seconds per liter, the appearance current (mmol/s) is administered into C which can reach a maximum value determined by the value of U_{in} and the resistors R_1 and R_2.

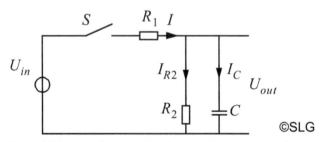

Figure 3.11   Electrical model equivalent of appearance and decay.

According to this model, R_1 represents the resistance for charge into the blood compartment capacitor C.

A relative large value of R_1 results in a measurable delay of the appearance level.

R_2 is a modeling equivalent for the amount of metabolic activity which is represented by the value of R_2.

A relatively high value of R_2 results in a low decay rate of the level to homeostatic values. Vice versa a low value of R_2 results in a relative quick decay of the output level.

$$U_{C\,max} = \frac{R_2 U_{IN}}{R_1 + R_2}\,(\mathrm{mol}/L) \qquad (3.38)$$

The resistor R_2 is the "bleeding" or leaking resistor representing the summation of all usage and metabolic processing effects.

The signal U_{in} can either be a constant level as a step function or be a gradual increasing level of the source.

The signal U_{out} stands for the total resulting level at a certain moment.

To analyze the dynamic behavior of the appearance and decay process, we will use analysis methods commonly used in electrical network theory.

This means that the laws of Ohm and Kirchhoff as well as the reciprocal principle and the superposition principle are valid.

The Laplace transformed time function will directly be derived from the electrical network and represents the network transfer function with complex variable $p = j\omega$.

The general transfer function in Laplace transform of the differential equation of this model can be written as:

$$U_{out} = \frac{R_2 U_{in}}{p R_1 R_2 C + R_1 + R_2} \qquad (3.39)$$

The uptake time constant τ_U is denoted as

$$\tau_{\text{appearance}} = \frac{R_1 R_2 C}{R_1 + R_2} \qquad (3.40)$$

The appearance time constant τ_D as

$$\tau_{DECAY} = R_2 C \qquad (3.41)$$

This implies that the uptake time constant is smaller than the decay time constant.

3.7 Appearance Time Constant

The appearance characteristic is the time domain solution of the differential equation that describes the voltage rise of the capacitor as a function of time.

In this example the input signal for the R–C network of Figure 3.11 is a step function. That means that at $t = 0$ the input signal rises very steeply from zero to a defined saturation value A. In our example the end value equals five.

In Figure 3.12 the step response on the capacitor of the network is shown.

The first approach will be discussed as the exponential model. The following calculations will be performed in the time domain.

Instead of the Laplace transformed expressions, the time domain solution of the characteristic differential equation for the transfer from U_{in} to U_{out} will be used for the determination of the appearance model parameters A and α.

$$U_{OUT} = A\{1 - \exp(-\alpha x)\} \tag{3.42}$$

In the example of Figure 3.12, we use $A = 8$ and $\alpha = 0.5$.

The value of a is found from the first derivative of the U_{out} time function in the point $\{0,0\}$.

$$\frac{dU_{OUT}}{dx} = 4\exp(-0.5x) \tag{3.43}$$

$A\alpha = 1.25$. Thus, $\alpha = 0.5$ and

$$\tau = \frac{1}{\alpha} = 2 \tag{3.44}$$

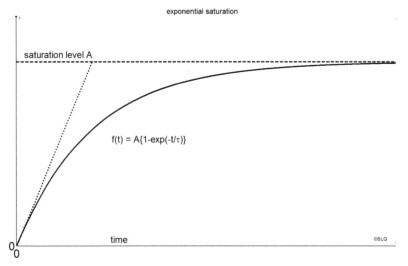

exponential saturation

saturation level A

f(t) = A{1-exp(-t/τ)}

time

©SLG

Figure 3.12   Time function of the appearance process and the expression of the first derivative.

In the point $\{0,0\}$ we find the tangent at the function U_{out} which intersects with the asymptote of $y = 5$ from which we find that $1/\alpha = 4$.

From the individual appearance response the belonging time constant can be obtained.

$$\tau_{appearance} = \frac{R_1 R_2 C}{R_1 + R_2} = 4 \tag{3.45}$$

3.8 Determining the Value of the Appearance Time Constant

Because the appearance function has saturation characteristics, such a function has been described by a Michaelis Menten function in the form of

$$y = \frac{Ax}{b + x} \tag{3.46}$$

Also the hyperbolic tangent function $tgh(x)$ has been applied.

Because in biological systems we will never encounter hyperbolic functions, discontinuities, or sharp cutoffs, but only exponential relationships as natural general solutions of non-linear continuous differential equations, we will use the solution of the differential equation that describes the impulse and linear response on the network of Figure 3.11 resulting in the exponential expression

$$y = A\{1 - \exp(-\alpha x)\} \tag{3.47}$$

From this form we are in the first place interested in the value of α because the reciprocal value represents the uptake time constant τ_U.

Dividing $\{2\}$ by the saturation value A, we need only one single point to calculate the value of α.

When the measured point is represented by (x_1, y_1), we find

$$\alpha = \frac{1}{x_1} \ln\left(\frac{1}{1 - y_1}\right) \tag{3.48}$$

This method is valid for the elementary function as denoted by $\{2\}$.

For the appearance and decay functions we have to take the homeostatic value into account by subtracting the value of the homeostatic level from the measured concentrations when a curve fitting procedure is carried out.

Another method to determine the complete appearance function is the use of two points of the appearance function (3.42) measured from a graphical presentation.

We can calculate the values of α and A by the determination of the first derivative of the function in these points.

Because the homeostatic constant does not play any role here, the subtraction has to be omitted.

For simplification we substitute $U_{out} = y$.

Measured coordinates are

$$P_1 = (x_1, y_1) \text{ and } P_2 = (x_2, y_2) \tag{3.49}$$

From

$$\frac{dy}{dx} = y' = \alpha A \exp(-\alpha x) \tag{3.50}$$

we find

$$y'_1 = \alpha A \exp(-\alpha x_1) \tag{3.51}$$

and

$$y'_2 = \alpha A \exp(-\alpha x_2) \tag{3.52}$$

Then

$$\frac{y'_2}{y'_1} = \frac{\alpha A \exp(-\alpha x_1)}{\alpha A \exp(-\alpha x_2)} = \frac{\exp(-\alpha x_1)}{\exp(-\alpha x_2)} = \exp\{\alpha(x_2 - x_1)\} \tag{3.53}$$

or

$$\alpha = \frac{1}{x_2 - x_1} \ln\left(\frac{y'_1}{y'_2}\right) \tag{3.54}$$

and with

$$y'_1 = \alpha A \exp(-\alpha x_1) \tag{3.55}$$

we find

$$A = \frac{y'_1}{\alpha \exp(-\alpha x_1)} = \frac{y'_2}{\alpha \exp(-\alpha x_2)} \tag{3.56}$$

and thus

$$\tau_{\text{appearance}} = \frac{R_1 R_2 C}{R_1 + R_2} = \frac{1}{\alpha} \tag{3.57}$$

The appearance curve has been characterized in this way.

3.9 Discussion

From the results of experiment 1, we learned that just by means of careful observation and careful measurements, we can register the flow results of a process. The word "results" means that we have recorded an equilibrium or homeostatic state of the experiment at a certain time. We were not able to monitor the details of everything that was happening during the transition from level mark 9 to level mark 8, etc. of the water container experiment. The transitional process between the measured points will be designated as *transient behavior* for future *dynamic* modeling experiments. Therefore, we took a sample, or a measurement at a defined time resulting in a sampled set of data. Notably, experiment 1 has been constructed with a container of which we know the dimensions by design as is for the amount of water we filled it with and the drainage opening at the bottom. From the plotted sampled measured results, we can find a suitable mathematical expression with a curve-fitting procedure that can be fitted on the measured data. In many cases, we can expect an exponential relationship between variables when we evaluate the final state in an equilibrium situation. The example of the water container is a suitable model for many situations in human physiology but it is not completely identical to the human body. It remains a model that shows a behavior similar to observations of equilibrium from physiology!

The electrical network is constructed and designed with well-known and characterized network elements, like the voltage source, the resistor, and the capacitor and is based on a well-established and validated electrical network theory. From the measurements of the behavior of the water container, we can derive a similar or even an identical behavior using the electrical network model. Electrical network models are very flexible and easy to simulate in a suitable environment with appropriate computer simulation tools.

Because the capacitor value and the resistor value can vary over a wide range, we can only derive a general parameterized model to fit an individual case where the measurements are taken from.

The moment we have characterized the specifics of a certain part of the physiological process, we can translate the results back to physical parts of the physiology. The application of electrical network equivalents will also be used in other examples of modeling.

3.10 Conclusion

Because the physiology of living entities and especially the physiology of the human body is so complex, it is difficult to carry out an accurate analysis of parts of the system, let alone the system as a whole. We can analyze all kinds of substances in the blood, hormones, electrolyte levels, etc. The biochemical analysis forms a knowledge base in modeling. The difficulties begin when we want to find out how the system parts actually work and what the relevant relationships of the belonging components are. When a model is identified, we can use the knowledge of the model structure to compensate for measurement noise and correct for possible outliers. In such a case, the identified model can be used as noise filter. This issue will be discussed in more detail later on.

From systems theory [2], we know that a closed-loop feedback system, like the HPT system, cannot be analyzed from the outside with relatively small disturbances or alterations in the same order of magnitude as the signal variations which could be part of the normal operation. When such a closed loop is opened, for example when the thyroid action is no longer present, we can analyze the dose response behavior of the hypothalamus–pituitary unit without the corrective actions normally encountered in a closed feedback loop. Only the behavioral responses can be analyzed, but not all the underlying influences and transients governed by numerous process and sub-process variables and parameters. Our modeling methods in the HPT system will be confined to the effects we can observe together with the common measurements of TSH, FT4, and FT3. From there we can build a relevant and verifiable theoretical framework.

References

[1] Johansen, J. (2014). *Graph 4.4.2: A Graphical Mathematical Tool*. Available at: http://www.padowan.dk/graph/
[2] Aström, K. J., and Murray, R. M. (2009). *Feed Back Systems*. Princeton, NJ: Princeton University Press.
[3] Hook, J. R., and Hall, H. E. (1991). Solid State Physics second edition, Wiley & Sons Ltd, 9780471928058

4

Medical Statistics and Mathematical Modeling Make Strange Bedfellows

"There are three types of lies – lies, damn lies, and statistics."

–Benjamin Disraeli (1804–1881)

4.1 Introduction

Originally, statistical methods and probability theory were employed to study stochastic phenomena with an element of randomness like classical molecular collisions in gases and quantum theory. Stochastic events emerge as dependent and independent events and can be structurally applied in nuclear physics, thermodynamics, and other related physical events consisting of many identical objects in a closed system.

Today we encounter statistical methods being used in the investigation of phenomena in the soft sciences such as psychology, sociology, anthropology, political sciences, epidemiology, economics, and many sub-disciplines of biology and medicine [1]. One of the greatest fallacies of statistics is the erroneous application of the P-value. This emerges from probability theory and is used in inferential statistics to evaluate a null versus an alternative hypothesis related to a given dataset. The P-value has to be associated with the model identification and model parameters used in the investigation. Without a validated model the use of the "P-value" has no meaning.

For example, to understand the variations in speed of a certain object, the measurements have to comply with the theoretical model that determines the boundaries for expected maximal deviations in such an experiment. According to general relativity, the speed of light cannot be exceeded.

Such a model, or theory, has to be developed upfront before measurements can be used for the fine-tuning of the model. When any measurement crosses beyond the theoretical boundaries, the result is an outlier.

In fields of so-called "inexact sciences" such as biological and social sciences, the *P*-value is typically used to test a null hypothesis against an alternative hypothesis. Neither of these hypotheses is validated with repeatable experiments. In this context, the *P*-value is usually interpreted wrongly as the probability that the null hypothesis is true. It is a supposed measure of the strength of the evidence against the null hypothesis in favor of the alternate hypothesis. Notably, one needs to keep in mind that what is statistically significant may not have any clinical or practical significance. Hence, it is recommended that working with notions as a *P*-value should be eliminated and discarded.

4.2 Pitfalls in the Application of Inferential Statistical Methods

If progress in clinical thyroid research has been stifled, a contributory reason lies in the presentation and misinterpretation of measured data of thousands of patients in a scatter plot. This kind of data presentation is commonly accepted in research on populations and aims to generalize using one common model to fit all individuals in that population.

In this respect, we refer to the scatter plots of thyroid function tests (TFTs) of multitudes of different, unrelated, and independent individuals often coming from different laboratories. Because every clinical laboratory uses its own standards with regard to the determination of concentrations in blood samples, generally the results of these labs cannot be mutually compared. One of the erroneous arguments proposed to justify such an approach is that errors are minimized and accuracy maximized by acquiring as many data points the researchers could lay their hand on.

A TFT of a healthy person reflects the normal operating equilibrium of thyroid stimulating hormone levels [TSH] and free thyroxine [FT4] levels in a single individual and is presented as a combined test result reflected as a coordinate in a two dimensional [FT4]-[TSH] plane [2] with an example given in Figure 4.1. Faced with a group of data points, this mass of accumulated results can be handled in three different ways. The first is simply with descriptive statistics to figure out the positions and dispersion of the measurements. The second is related to the inferential interpretation of the measured data in order to find or identify a certain relationship. The third is intended to find predefined similarities between objects.

Let us start with descriptive statistics presenting the measured data as an observation. The only value of this result is to show what has been encountered.

1. During the treatment of thyroid patients, TFTs are performed regularly for monitoring and therapeutic purposes in a certain population. The most common test is the determination of the serum TSH concentration [TSH], which has a definite clinical significance. Depending on the situation, the value of the serum FT4 concentration [FT4] will be determined. In the following, only a pair of [TSH] and [FT4] will be defined as a TFT and depicted as a point in the [FT4]-[TSH] plane. Applying descriptive statistics on a scatter plot of (TFTs), we get an idea about the range, the position, and the density of the measured results. This can provide insight into the expected distribution of TFTs and also indicates an area of the highest density.

However, we have to appreciate the fact that each TFT point as presented in the scatter plot has absolutely no direct relation to any other TFT point and none of these separate data points have any influence on each other! Take for instance this: the [TSH] value of person A has no relationship to the [FT4] value of person B, and neither has the [TSH] value of person B any influence on the [FT4] value of person A! The only commonality of the TFT's is that they relate to a variety of the human physiology of which the physiological realm is known at the level of an individual human being. A similar situation can be found in the frequency distribution of body length, weight, or blood pressure.

Because the TFTs are matched coupled values of [FT4] and [TSH], we can obtain a picture in two dimensions resulting in the use of the bivariate ranges for [FT4] and [TSH]. This is indicated in Figure 4.1.

2. In many cases, a scatter plot of TFTs taken from numerous individuals is considered to provide a cross-sectional model or description of the inter-individual relationship of [FT4] and [TSH]. The relationship is then "found" with a linear regression line drawn through the "data cloud" of TFTs by using an arbitrary fitting procedure.

However, a regression line, according to a theoretical functional relationship, is only meant to find the best way to describe the $x - y$ relationship

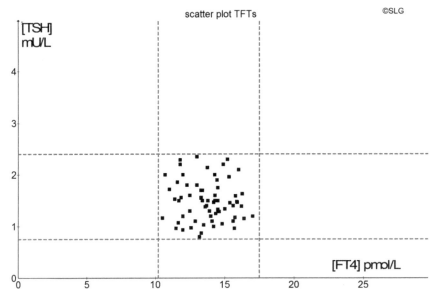

Figure 4.1   Scatter plot of measured thyroid function tests (TFTs) and observed bivariate ranges for [FT4] and [TSH]. The boundaries indicated by the dashed lines show roughly the area of interest.

of one and only one object referring to one and only one relational property of that object. In such a case, multiple measurements related to that single object will be investigated in order to identify a functional relationship. The result will then be a series of measured points that have a fundamental mutual relationship and these can be interconnected, from which a mathematical model, describing this relationship, can be identified and derived by means of the method of least squares [2].

In the TFT scatter plot as described above which represents measurements of a large group of different unrelated persons, we encounter a universe of different, unique, independent individuals from whom only one measured [FT4]-[TSH] data pair is available per person. In numerous papers where these misunderstandings of measurements were presented, there was never a mathematical explanation of the physiological significance and interpretation of the expected "model," nor a clear understanding of the measurement methodology. Commonly, the presented measured data were obtained from different laboratories with different [FT4] and [TSH] assays representing different reference ranges for [FT4]. This resulted then in unrelated data sets and unrelated measurement errors [3]. Also there was no explanation of the

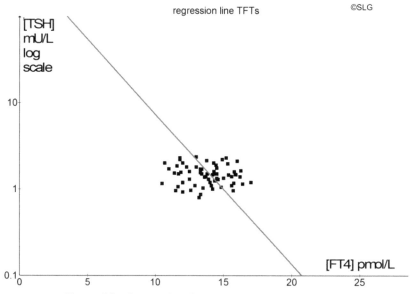

Figure 4.2 Scatter plot of TFTs with linear regression line.

significance of such a model, nor was there any theoretical prospect for an application.

In Figure 4.2 we see an example of a phenomenon that is too frequently encountered without any detailed notion of the underlying physiological and mathematical principles [4, 5].

This very common presentation is displayed in published research results from which one could draw the conclusion that the general model of the relationship between FT4 and TSH concentrations is represented by the linear regression plot with a logarithmic scale for [TSH] and a linear scale for [FT4]. This phenomenon is in the clinical domain known as log-linear relationship of TSH and FT4. Remarkably, if the scatter plot actually displays a symmetrical distribution when the test subjects are normal and healthy, a regression line should not be found in any direction. Then the question arises: "Why a regression line at all?" In what way does the TFT of person A have any relation to the TFT of a different person B? This is something like a fashion designer who establishes the average measurements for a suit and declares that the suit should fit everyone. This way of modeling will not lead to applicable results or theoretical developments.

A population of individuals will show some similarities in the sense that nobody of the adult population will exceed the length of 3 meters and nobody will be shorter than 0.5 meter. This is exactly what we learn from descriptive statistics, but it will never tell which person will have what length from the general presentation. We have to understand that the TFT result [6] of every person is a fundamentally independent entity with very distinctive physiological properties being the result of the individual's genetic makeup.

Therefore, we cannot identify any relationship between the measured TFTs as presented because they have no mutual relationship whatsoever, other than that these measured values are derived from humans.

4.2.1 Explanation of the Observed Log-linear Relationship between [TSH] and [FT4]

Still, the scatter plot presentation has led to the idea that the relationship between FT4 and TSH concentrations was logarithmic for the [TSH] scale and linear for the [FT4] scale. We can imagine that from evaluations of published data of thousands of measured [TSH]-[FT4] pairs, a number of persons had an almost identical individual [FT4]-[TSH] relationship, which will be discussed in one of the following chapters, but some of these individuals probably suffered from a thyroid disorder. In such a case, we can find groups of TFTs that can be identified to be part of a negative exponential-shaped curve belonging to these similar persons! A scatter plot of a large population would show a number of TFT groups ordered according to such an exponential-shaped curve with a linear scale for [TSH] or a straight line with a log [TSH] scale-linear [FT4] scale relationship. In early literature, this interpretation has led to the correct but unfounded conclusion that the relationship of log [TSH] versus linear scale [FT4] is a straight line.

In general, a scatter plot of measured results shows a distribution of the characteristics of many individuals which can be considered as a valuable asset of general information. The method and interpretation of clinical research has led to strange and confusing situations when the reference ranges for [FT4] and [TSH] were used to decide whether a thyroid patient was correctly treated with hormone replacement therapy or not. Research of the recent years [2, 3, 6] has shown that every individual has a very distinct and narrow range for the optimal value of [FT4]. This subject will be discussed in detail in Chapter 10. Although the values of [TSH] can be measured with high sensitivity (sometimes to three decimal places), the actual relative accuracy is at best 5% if the TFT of a patient has been performed on a blood sample taken on a specific defined time of day, because the value of [TSH] varies according to an individual circadian rhythm [7]. For a person treated with synthetic thyroxine, or levothyroxine (L-T4), the blood test should ideally be done in the morning between 7:00 and 9:00 h before taking the levothyroxine medication and breakfast. This sampling method guarantees a consistent time of day and the minimum disturbance in the blood pattern of the patient. Depending on the time of day, the value of [TSH] can vary by a factor of two as described in earlier publications [7]. As mentioned before, [TSH] is easy to measure and is very sensitive to all kinds of conditions. However, we have to keep in mind that the [TSH] is the signal used by the normal hypothalamus–pituitary–thyroid (HPT) loop to establish a stable value of [FT4] which is the target control variable.

All these experiments with data from thousands of people have only provided some notion about what could be considered as "normal" for the average person. In clinical practice, the most commonly used reference range for [FT4] is $10 < $ [FT4] $ < 20$ pmol/L and for [TSH] is $0.4 < $ [TSH] $ < 4$ mU/L. The fact that person A is well with an [FT4] = 12.5 pmol/L and a belonging [TSH] = 1.85 mU/L cannot be compared with the set point of another person B with an [FT4] = 16.2 pmol/L and a belonging [TSH] = 2.35 mU/L.

These situations are actually very realistic examples observed from a study of 70 persons with the model we will discuss later. We established the validity of these set points [8].

A clinician could, based on the routinely applied reference ranges, make a judgment that when person A had a TFT of [FT4] = 15.8 pmol/L and a belonging [TSH] = 0.85 mU/L and for person B a [FT4] = 13.3 pmol/L and a [TSH] = 3.5 mU/L, both persons are probably healthy as their TFTs are well within the reference range. Reference ranges result from descriptive statistical presentations and have an indicative value for a first diagnostics analysis, but cannot be used as methods for the individualized treatment of patients.

4.3 Abuse of Mathematical Models

4.3.1 Critical Notes

The design and use of mathematical models also have their pitfalls and drawbacks. In most cases of biological modeling, the modelers have been educated in a certain restricted way, as if biological or medical modeling of the endocrine system would be something other than technological modeling. Actually, it is exactly the same. During a review exercise by network theorists, a number of biological and medical models of the HPT, described with the most modern mathematical methods, were assessed. The common result and conclusion was that none of the models had been validated or tested with real data, nor were any of the models self-identifiable because of numerous not estimable model parameters.

In practically all papers on mathematical modeling, the significance of measurements related to the subject are not present or ignored and at best used as a vague kind of indication. The model is generally introduced as "let us consider the following equation," without the argument or explanation what such an equation could represent.

Normally, the expansion of non-subject-related mathematics results in a fuzzy, but mathematically correct discourse. In order to compensate for practical necessities to get the model to fit the data to a certain degree, numerous not estimable model parameters have to be introduced. However, we have to realize that many internal parameters and variables in the endocrine system cannot be observed or directly measured. For that reason, the introduced model parameters have to be "estimated" with a source of uncertainty. The mathematical method looks very solid and impressive, but the development and expansion of these mathematics were not related to the subject of investigation and ended up in a non-compliant and non-transparent expose [8–10].

A usable model has a minimum of parameters and is self-identifiable, meaning that the model parameters are identified and extracted from the measured data [11]. Another common flaw in mathematical modeling is the inability to come to a clear conclusion based on belonging measurements (if there were any) and is in the discussion normally clouded by arguments of numerous model parameters to be assessed or estimated. Even worse, when the calculated result had no quantitative relationship with the measurements, the results were hidden in a blurred wide band of graphical data presentations, or hidden in a statistical presentation indicated in a figure, representing possible outcomes [7].

The finding of the mathematical modeling papers assessment was that, although the examiners were not quite familiar with the applied terminology, the results were found to be very fuzzy or even not presented and in almost all cases a validation was lacking.

Sometimes, as far as there were results, the mathematical form was so highly expanded and complex that the only way left to verify the outcome with data was the attempt to put the result in a simulator and see what happens. None of these modeling publications had provided any clarity or insight, let alone any practical or theoretical value [8–10]. None of the models were identifiable with measured data nor were they translatable to the underlying physiology. Furthermore, in many of the mathematically oriented biology journals, the level of mathematical complexity is considered to be the leading qualifier to determine whether a paper will be accepted for publication or not.

This is what we could call the pitfalls and the abuse of complicated fantasy mathematics to impress and intimidate readers in order to hide ignorance and incompetence. It is nice to have some knowledge of non-linear or partial differential equations, but if one does not know to what it applies or to what phenomenon it is related, one cannot validate or test the result with real data.

4.4 Future Developments

The following might provide a better example. Instead of statistical methods, the structural application of mathematics for the modeling of biological systems can be used in a very successful and transparent way. We can also look at the successes in physics where fantasy and non-relevance is immediately criticized and eliminated when the model cannot experimentally show what it has promised or is supposed to prove. As a dramatic example, we can mention the cold fusion of hydrogen reported by Pons and Fleischmann in 1989 [12]. Also, theoretical work on the string theory seems not to be verifiable by means of experiments.

In technology, this can have dramatic effects, especially when lives are at stake. Think about safety in vehicles, airplanes, and space craft. Also modeling related to safety issues in a hospital environment has direct consequences when patients would perish because of faulty modeled and programmed equipment, like monitors and alarms in intensive care and X-ray equipment running out of control.

We have noticed that the biological system modeler in the biological modeling literature does not run any risk when the modeling activities are only related to abstract or non-direct human-related subjects.

Nobody will perish, almost no one is really interested, and no one has taken the trouble to verify the published result or check on the theory by means of an experiment. In the world of medical and biological modelers, everyone can afford to publish any theory or whatever non-verifiable mathematical fantasy. This will be fundamentally different the moment that the health and safety of people are involved and the researchers are made accountable for what they have published or manufactured.

Then the question arises:

In what field is mathematical modeling successful? As mentioned earlier, mathematical modeling is particularly useful when certain hypotheses cannot be proved with human experiments because of ethical restrictions, or when there are repercussions for experimental failures, when safety issues are involved, or large amounts of money can be lost when a mistake is made, or even worse, when lives are at stake. On the other hand, the rewards of a correct model can be tremendous in the form of progress in understanding the biological functions.

A compelling example applies especially to the semiconductor industry where the modeling, characterization, and testability of the integrated circuit technology are of paramount importance for a final successful product. One mistake in the semiconductor model can immediately result in a loss of millions of US dollars due to non-functional designs. As a follow-up for safety controlling electronic systems and space avionics, formal structural test and verification methods are of unparalleled importance. These examples could be an inspiration for modeling successes and responsibility in the medical field.

4.5 Conclusion

For statistical methods in research, we can conclude that this approach is relatively futile and even wrong compared to statistical mechanics, quantum mechanics, plasma physics, or any other established mathematically founded formal theoretical framework. Statistical analysis can be applied only to identical objects in a closed system in which a functional relationship between the objects of interest has been identified. This ensures that the statistical analysis is confined to variables with an identified mutual relationship in a defined isolated environment.

This is clearly not the case when we investigate human disorders of cross-sectional populations because every person is fundamentally unique and independent! Statistical methods as currently in use in the liberal arts and medicine cannot provide a theory or understanding about the subject of investigation and the related investigated data. Also, the display of nonsensical mathematical modeling as published in so-called advanced mathematical biological and medical journals has often not delivered any useful result other than impressive and useless mathematics. On the other hand, mathematical modeling is meaningful only when it applies to identifiable problems and can be understood in terms of a relevant physiology and is validated with belonging measurements.

When the stakes, risks, and responsibilities are high enough, the modeler should be held responsible and accountable for the results as is commonly the verdict in physics and other related science disciplines [12]. In the following chapters, a well-founded and validated theoretical framework will be discussed.

References

[1] Strasak, A. M., Zaman, Q., Pfeiffer, K. P., Göbel, G., and Ulmer, H. (2007). Statistical errors in medical research-a review of common pitfalls. *Swiss Med. Wkly.* 137, 44–49.

[2] Goede, S. L., Leow, M. K., Smit, J. W., and Dietrich, J. W. (2014). A novel minimal mathematical model of the hypothalamus-pituitary-thyroid axis validated for individualized clinical applications. *Math. Biosci.* 249, 1–7. doi: 10.1016/j.mbs.2014.01.001

[3] Goede, S. L., and Leow, M. K. (2013). General error analysis in the relationship between free thyroxine and thyrotropin and its clinical relevance. *Comput. Math. Methods Med.* 2013:831275. doi: 10.1155/2013/831275

[4] Rothacker, K. M., Brown, S. J., Hadlow, N. C., Wardrop, R., and Walsh, J. P. (2016). Reconciling the log-linear and non-log-linear nature of the TSH-free T4 relationship: intra-individual analysis of a large population. *J. Clin. Endocrinol. Metab.* 101, 1151–1158. doi: 10.1210/jc.2015-4011

[5] Hoermann, R., Eckl, W., Hoermann, C., and Larisch, R. (2010). Complex relationship between free thyroxine and TSH in the regulation of thyroid function. *Eur. J. Endocrinol.* 162, 1123–1129.

[6] Leow, M. K., and Goede, S. L. (2014). The homeostatic set point of the hypothalamus-pituitary-thyroid axis-maximum curvature theory for personalized euthyroid targets. *Theor. Biol. Med. Model.* 11:35. doi: 10.1186/1742-4682-11-35

[7] Roelfsema, F., and Veldhuis, J. D. (2013). Thyrotropin secretion patterns in health and disease. *Endocr. Rev.* 34, 619–657. doi: 10.1210/er.2012-1076

[8] Eisenberg, M., Samuels, M., and DiStefano, J. J. (2008). Extensions, validation, and clinical applications of a feedback control system simulator of the hypothalamo-pituitary-thyroid axis. *Thyroid* 18, 1071–1085. doi: 10.1089/thy.2007.0388

[9] Distefano, J. J., T.. Kitchener, T., Wilson, C., Jang, M., and Mak, P. H. (1975). Identification of the dynamics of thyroid hormone metabolism. *Automatica* 11, 149–159.

[10] Pandiyan, B., Merrill, S. J., and Benvenga, S. (2013). A patient-specific model of the negative-feedback control of the hypothalamus-pituitary-thyroid (HPT) axis in autoimmune (Hashimoto's) thyroiditis. *Math. Med. Biol.* 31, 226–258. doi: 10.1093/imammb/dqt005

[11] Middelhoek, M. G. (1992). *The Identification of Analytical Device Models*. Ph.D thesis, Delft University Press, Delft.

[12] Fleischmann, M., and Pons, S. (1989). Electrochemically induced nuclear fusion of deuterium. *J. Electroanal. Chem. Interfacial Electrochem.* 261, 301–308.

5

Systems Theory Applied on the Modeling of the HPT Axis and First Principles of Feedback and Homeostasis

"The laws of Nature are but the mathematical thoughts of God."

–Euclid (325–265 BC)

5.1 Introduction

Modeling in biological systems is in many cases based on the notion that biological processes can be described as time-dependent processes.

It is a valid observation that state variations in a system in general are developing processes as a function of time. However, in biological systems, the time-dependent dynamics of various functional parts are intertwined with numerous variables and influencing circumstances that render this approach to process analysis a mere futile exercise.

In reality the time-dependent dynamics are too complex to come to a satisfying modeling result based on the approach that the internal observable parts could be described by differential equations. This becomes obvious when we realize that the various processes, biological units like organs, etc., are operating on a different time scale, with different time constants and the observable parts and processes are limited to a few. Furthermore, biological modeling in most cases aims at a generalized description of a certain system or a phenomenon which is impossible because of the tremendous diversity in bio system characteristics and individuals.

Therefore, this method will not lead to a successful model.

Another more important property of a biological system can be described by the conditions of a defined equilibrium. Here we can make the distinction between dynamics as a result of a sudden change in conditions, acceleration, resulting in a transient situation and the state maintained over a longer period, continuous speed, that we can identify as a dynamic equilibrium or homeostasis.

The use of computing power and analysis can potentially retract researchers in some cases from the solid fundamental understanding about the subjects they are investigating.

The lack of insight is then masked by using the computer with the possibilities of simulation without understanding the models that are part of the simulation environment. Here we can make use of the extensive experience from electronic design practices of electronic engineering based on electrical network theory.

From electrical network theory we appreciate that these networks are physically existent and can be realized by means of electrical components. Thus, when we can find a similarity of the electrical network response and an observed physiological response we have the possibility to translate the physiology into an electrical network model.

There are many examples in electronic engineering where the designer of an analog circuit was not aware of the physical limitations of the transistor models he was using. In such a situation we could find simulator results far beyond the reality of normal physics, because a computer will not overheat by the presentation of megawatt results in a tiny transistor!

This example illustrates the fundamental need to validate the simulation models with a realistic test environment. This implies that all modeling should be validated with real measurement results of the modeled subject. We also have to keep in mind that a real model in hardware is always superior to any computational simulation; however, sometimes, a computer model is the only option for investigation.

There is an alternative with which the final results of complex systems can be analyzed in the form of states of dynamic equilibrium. Apart from the equilibrium model, the transient behavior can be modeled by means of stimulus and response. The investigated subject is then considered as a black box.

Another point of interest is to investigate the generalization of such dynamic behavior. This means that the analysis of a dynamic process only serves a purpose when we are confronted with corresponding timescales of the subject we are investigating, which occurs in transient phenomena after a change in the initial homeostatic conditions.

The main goal of modeling is to understand behavior and to be able to provide reliable predictions under defined circumstances. When a biological part or module can be modeled and analyzed in such a way, we can use the validated models to synthesize more complex systems.

In most cases biological models are non-linear, but in all cases continuous and analytic over the dynamic interval of interest. The input parameter space is limited to definable extends.

An illustrative example of dynamic behavior can be found in the diffusion process of a tiny ink drop in a glass of water.

The moment a small drop of ink is added to a glass of clear water at room temperature, we see that the drop is dissipating in the turmoil of the surrounding moving water molecules (Brownian motion) and after half an hour, the water is uniformly colored by the ink.

The diffusion process after the input of the ink drop can be inspected visually and we see the drop spreading over the water volume as a function of time.

The dissipation expands at an increasingly slower, negative exponential rate when the water volume would be without boundaries.

We can wonder what we would accomplish to analyze this diffusion process.

This is only of interest if we could influence or stop this, or freeze the process, at a certain defined amount of time and use the result that has been predicted by the equations of the diffusion model. Note that we cannot reverse such a process because of the second law of thermodynamics.

This kind of diffusion process analysis has successfully been applied in the design of controlled diffusion profiles in semiconductor processes that had to comply with very precise dimensions and properties.

We could wonder what to do if a certain chemical process is already active and how we could influence it to our desires.

A very simple example is a wood fire that can be stopped by a direct cooling activity (water) and/or the deprivation of oxygen.

Such a process is slow enough to realize the required amount of influence to stop it.

High-speed processes are much more difficult to influence, like an explosion reaction.

The chemical reaction speed is in such a case is too high to influence in a practical manner.

However, when dynamic processes cannot or hardly be influenced during the occurrence of such a process, we can only try to find out in what way these occurrences influence transient behavior. Then we can analyze and model this transient.

In biological systems we know the effects of externally added substances as pharmacokinetics and dynamics.

In many cases we can predict the result of a certain amount of food, liquid, medicine, or chemical substance once it has been characterized for a certain biological process or entity. After administration of a substance, it is sometimes possible to intervene to prevent negative or lethal effects by means of a substance from which we know it can neutralize the initial effect.

This means that the dynamic processes in these cases have been analyzed and characterized related to initial starting conditions and the end result over a certain amount of time. It is practically impossible to obtain a generalized model for the belonging biodynamic processes because these are always individually determined.

The individualization of the model can be realized by means of parameterization and the investigation protocol dedicated to only one object.

In a second stage of research, the results of multiple objects can be compared.

A suitable approach for biological modeling can be realized with the comparison of states of equilibrium as the end result of a characterized dynamic process which is the equilibrium state of initial transients. The equilibrium or homeostatic states are stable and the end result of such dynamic processes.

Systems characterized by a state-dependent response are defined as state transfer functions which are considered as time independent and the small signal transfer function (small variations around a defined operating point on the transfer curve) is represented by the first derivative of the state transfer function or large signal transfer function.

Transient characteristics are described as time dependent and can be represented as an impulse, step, ramp, periodic, or any other analytic function of time.

In biological systems the time dependency is determined by the type of manifestation and the belonging time constants.

For example, the thyroid system is characterized by large time constants for the dominant equilibrium situations (4–6 weeks) while the adrenal system is typically characterized in the second to millisecond time frame.

In many cases it is possible to find a suitable model for the relationship between these homeostatic end points and their initial input condition change.

In traditional clinical thyroidology, the level of serum [TSH] is the leading diagnostic indicator of the presence of any issues impeding the normal operation of the hypothalamus–pituitary–thyroid (HPT) axis. This is understandable because it is one of the first discovered control hormones that could be measured with an acceptable accuracy and repeatability. The measuring methods (i.e., assays) for the detection of [TSH] have undergone tremendous improvement over the years notably with respect to the sensitivity as denoted successively by first-, second-, and third-generation assays. Ironically, despite the high detection sensitivity, the precision is currently not better than plus or minus 5%.

When we encounter [TSH] values higher than 10 mU/L, the standard laboratory measured result reports it with an accuracy of two or even three decimals, while on the other hand the most important variable, the value of [FT4], is often rounded off or even truncated to a crude integer value! Just imagine that the natural diurnal variations in TSH levels can be found between 1 and 3 mU/L over 24 h! [1] In this light, the accuracy and sensitivity seems pointless [2].

In a healthy person, the thyroid gland is part of a control system that maintains the specific individual level of serum free thyroxine [FT4] at a specific value corresponding to what we will define as the set point value of [FT4] for the HPT system. The homeostatic set point is expressed as a combination of [FT4] and [TSH] values which is specific to any individual like a finger print [3–5]. This set point value of [FT4] is maintained robustly and efficiently via small control variations of [TSH]. As mentioned earlier, this specific value of [FT4] occurs commonly within the [FT4] population reference range $10 < [FT4] < 20$ pmol/L. Similar reference range values for the [TSH] are defined according to $0.4 < [TSH] < 4$ mU/L.

However, we have to keep in mind that the main target for the HPT system is to keep the specific individual's set point [FT4] value tightly defended. Because the functioning thyroid is an integral part of a closed feedback loop to lock the [FT4]-[TSH] coupled hormones into the set point uniquely optimal for an individual's normal euthyroid state, it is inherently difficult to tease out the causal relationships and effects governing the control of the HPT axis. A closed-loop feedback system of a normal HPT cannot be analyzed as any signal will be canceled out by the control mechanism! In this sense, a signal is a concentration value of which the amplitude is either a time-dependent or quasi-static input variable. However, when the variations of concentrations are relatively slow over long time periods, the description of the system behavior can be done in terms of characteristic variable dependencies. In such a case, the system behavior is described in a non-time-dependent or quasi-static manner and we only observe the new stable homeostatic situation or system state. This means that when we enhance externally the amount of [T4] by means of L-T4 medication at a certain time, in a normal healthy person, we will only encounter a temporarily (4 h) rise of [FT4] after which the closed loop will regulate this excess amount back to normal values. Although this disturbance can be interpreted as a transient, which it really is, the transient time aspects are not directly of interest here. The transient is then regarded as a short disturbance, causing temporary "turmoil and turbulence." When the "turbulence clouds" have settled, we examine the new condition or equilibrium state of the system. This state represents a stable condition valid over a relative longer time period. As an example, we can consider the value of [FT4] as a stable one for a period longer than a week.

Keep in mind that the occurrence and maintenance of a certain thyroid state can be modeled as a flow of [T4] and [T3] "currents" from the thyroid and their control of [TSH] current from the hypothalamus–pituitary (HP) system. These current flows are represented by the variable's time-dependent pharmacokinetics and pharmacodynamics. However, when a certain current is more or less constant over a certain period of time, we can define this equilibrium result as a stable state. Still this state is maintained by a constant inflow of signal, but finds its equilibrium by an equal outflow of the signal in question. As such, time dependencies are then not an issue and all modeling and calculations can be performed with normal algebraic equations. This idea of equilibrium has been modeled with the example of a leaking fluid container with a fixed area of the outflow opening or a similar electrical capacitor with a fixed leaking resistor as described in Chapter 3. This model concept will be frequently used and is generally applicable to all integrating biological functions and homeostatic conditions and even more important, translatable to the related physiological behavior.

When we are interested in the time dependencies of a system transient by a strong disturbance in the HPT loop, we get insight into the transient behavior though such a test signal.

This will be discussed in the dynamics of the HPT feedback characterization. In some test situations, this can be applied to a human being without inducing harm. However, the quasi-static investigation of system properties is best applicable to the HPT loop because of the relative long half-life of thyroxine. In contrast, when time dependencies of concentration variations are more relevant, such as those encountered in systems of glucose regulation, hypothalamus–pituitary–adrenal system, and brain functions, the analysis of time-dependent concentration variations and bio-electrical signals predominate because such perturbations critically dictate the system's behavior and therefore have practical and clinical importance. Similar situations of the time dependencies of HPT hormone concentrations are described and analyzed in the chapter devoted to the [T4] and [T3] pharmacokinetics and pharmacodynamics.

In the early years of the steam engine development, mechanical engineers needed stabilization methods and devices to control the rotations per minute (RPM) of a steam engine. Controllers or governors were used on train locomotives, pumps, and electrical power generators to maintain a predefined RPM in order to provide a controlled speed or constant voltage and frequency level. The realization of these controllers was rather crude and completely mechanical. The stabilization mechanisms were only understood on an intuitive level. However, these regulators show great similarity with biological homeostatic control systems [6].

Another well-known example of a feedback-controlled system is the central heating system in a house, room, or a building. The thermostat with a temperature sensor is set to a certain desired temperature. Depending on the set value of the thermostat ambient temperature, the thermostat will send a message to the heating device to enhance the water temperature in the heating radiators. A temperature decrease will occur with heat loss, defined by the room's thermal time constant. The time dependency of temperature drop in a room when the heater is turned off is exactly the same as the decay behavior of FT4 concentrations in a person without an operational thyroid when the L-T4 medication has stopped. This decay is a negative exponential function as will be discussed later in detail.

The HPT feedback system is quite comparable to a central heating system. The thermostat function is represented by the HP unit measuring the ambient amount of [FT4] (analogous to temperature). Internally, the HP regulator has a natural physiological reference [4] that provides a decision criterion to trigger the production of more or less [FT4] by means of more or less [TSH] signaling, which will be followed by appropriate responses of the thyroid. An enhanced [TSH] input on the TSH receptors of the thyroid follicular cells results in an increased amount of [T4] (+ some T3) thyroidal output and vice versa. Thyroid-generated T4 will be rapidly bound by protein binding which results in a very small fraction of bioactive free T4 (i.e., [FT4]).

The level of ambient [T4] and [FT4] will only drop as a result of an absence or decline of newly produced T4 as pharmacokinetically determined by the half-life value of T4. In the simplest model idea, this compares well with the thermostat-controlled heating system. External factors such as some effects of a feedback control system can be observed. For instance, the administration of exogenous T4 in a tablet form called levothyroxine (L-T4) can be studied. According to this example, we can compare the temperature sensor of the central heating system with the HP which senses the ambient temperature ([FT4]) around the thermostat itself.

In 1927, Harold Black, an electronics engineer at AT&T Bell Laboratories in Murray Hill, NJ, United States, worked on the development and improvement of telephone line amplifiers. In those days, the line amplifiers were built with electron valves (vacuum tubes) as active device for amplification. Because these valves were heated to generate electrons, the amplification and signal characteristics were not constant over a long period of time. This deterioration of the valve characteristics caused reliability problems and associated costs to replace faulty amplifiers in the field. Harold Black solved this problem by means of the negative feedback of the amplifier output to the input which provided stabilization and precise control of the amplification factor. The history and literature about feedback is much richer and more detailed than we can discuss here. In this chapter we will focus on relevant aspects of system theory and the various feedback mechanisms [6].

Feedback is here defined as the coupling of a portion of the output signal of a system to the input stage where the output signal is compared with the properties of the input signal. This is performed in such a way that the output signal is counteracting the moves of the input signal. The compared result will be put into the system chain again and processed to the output. Here we have a description of a closed loop that cannot be analyzed from the outside with normal signals. This enigmatic effect is caused by the properties of a closed-loop negative feedback system and is mainly due to the action of what we term the loop gain. The higher the loop gain, the more stable and accurate the system will maintain the reference set value. An overview of feedback systems is written by Astrom and Murray [6].

Let us now visit the relevant system definitions in order to elaborate our modeling with a mathematical strategy.

5.2 Definition of System Components

In the following, we introduce the definitions of signals, function blocks, and comparators. The input and output variables of a system function block are defined here as signals. These signals can show all kinds of variations as a function of time. In order to avoid unnecessary mathematical manipulations with differential equations, the signals will be defined as static input and output levels or states that can be moved from their static position. The notion of a signal can also be interpreted as a time-dependent displacement. This leads to a signal level factor (amplitude) indicated by a letter X, Y, etc.,

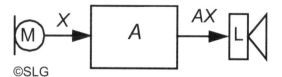

Figure 5.1   Basic amplifier configuration.

representing either a static position or an amount of maximum displacement. A signal can be considered as a state, deviating from a previous one with a deviating speed varying from zero to a certain defined amount. To give a simple yet very common example, we use an audio amplifier shown in a block diagram which is representative for all other block diagrams to be discussed.

In Figure 5.1, we show the block diagram of a basic audio amplifier. Imagine that X is an electric microphone signal, representing a continuous harmonic mechanical displacement as a function of time, which is amplified by the audio amplifier with the factor "A" and results at the output in AX, also a time-dependent harmonic displacement, which is made audible by means of loudspeaker L, transforming the electrical displacement into a mechanical displacement. The input signal X multiplied by A results in the output signal AX. This multiplicative operation applies across all signal blocks.

In Figure 5.2, we see two examples of linear transfer functions of an amplifier. Signal transfer 1 is the black line $y = x$ resulting in a gain of 1. Variation of the input signal between 2 and 3 on the x-axis results in an output signal on the y-axis of 2–3. The amplitude of the input signal $= 1$ which results in an output amplitude of 1 implies that the signal gain is 1.

Signal transfer 2 is the dashed line with transfer function $y = 5x$. From the form of the function, we can see what the gain will be. An input signal on the x-axis varying from 0 to 1 with an amplitude of 1 will result in an output signal on the y-axis varying from 0 to 5 with an amplitude of 5.

The gain factor thus equals:

$$G = \frac{\text{output amplitude}}{\text{input amplitude}} = 5 \qquad (5.1)$$

When the transfer function is defined as $y = f(x)$, where y is a non-linear function of x, we can define the gain factor G in a defined point P of y as the relation

$$G = \frac{\Delta y}{\Delta x}, \qquad (5.2)$$

where Δy is the output result of a small variation Δx around the value of x_p.

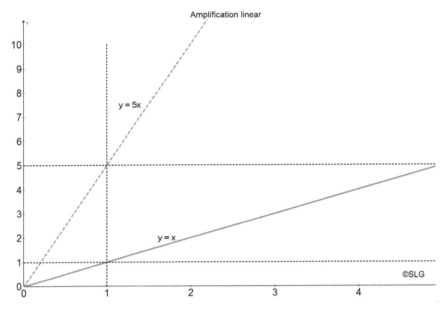

Figure 5.2   Signal gain of a linear amplifier.

These small variations may be interpreted as linear displacements on the x-axis and y-axis and therefore are considered to be linear in that confined area around the point of operation P. For infinitesimally small variations of Δx and Δy, we find the differential transfer factor by:

$$\lim_{\Delta x \to 0} \{G\} = \frac{dy}{dx} = \frac{df(x)}{dx} = f'(x) \tag{5.3}$$

In other words, we define the gain as the steepness of the function around a central point P of operation. This is illustrated in Figure 5.3.

In Figure 5.3, the input signal on the x-axis varies from 1.5 to 2.5, and the amplitude is 1. At point P on the transfer curve $y = x^2$, we determine the steepness by means the derivative of the function in P.

$$G = f'(x) = 2x \tag{5.4}$$

For the equation of the dashed line, we find

$$y = 4x - 4 \tag{5.5}$$

Thus, for input state $x = 2$, we have an output state of $y = 4$ and we find the gain factor $G = 2 \times 2 = 4$. This result of the output signal can be demonstrated

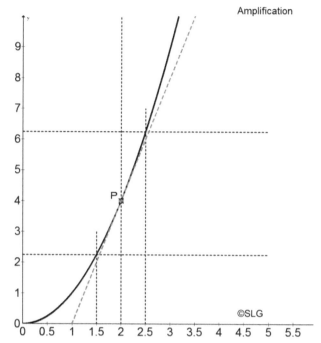

Figure 5.3   Signal gain of a quadratic amplifier around a chosen point P.

with the indicated output variations on the *y*-axis between 2.25 and 6.25, thus giving rise to an output state difference of 4. From this example, we learn how to calculate the differential transfer function of any form or non-linear shape just by the determination of the first derivative in a point of operation P central of the input variations. The signal difference is indicated as a different state indicated with a number that can be increased and decreased. In this way, the signal will not be seen as a continuous time-varying function. Based on this notion, we can easily analyze all kinds of transfer functions without the burden of unnecessary time functions or differential equations.

In the following, all signals will be indicated with a capital. In general, the signal is a composition of a static component and a time-varying part. Besides the static part, which can play a role under certain circumstances, we only focus on the time-varying component of the signal. In the terminology of linear systems theory, we denote this situation as is shown in Figure 5.1.

As mentioned in Figure 5.1, we encountered the notion of a transfer function in the form of an amplification or gain factor A. This amplification factor represents in the most elementary form a transfer function.

The amplification operation is defined here from very slow displacements in the order of centimeters per second to the highest imaginable movements like Tera-Hertz speeds of variations. In other words, there are no speed limits.

In most transfer functions, the letter A represents the gain factor from direct current to infinite frequencies. In practical systems, the speeds of signals are generally limited. We then encounter a signal speed limitation defined by the highest frequency that will be normally amplified. At higher frequencies there will be no significant measurable output response anymore. Depending on the transfer function properties, we encounter one or more frequencies where the signal response will be decreased which we will define as so-called poles. The explanation of a pole can be found in more detail in textbooks about complex variables like "Spiegel, complex variables" in the well-known Schaum series of physics and mathematics and the application is well explained in feedback systems written by Astrom and Murray [6]. For our definition of a pole, we expand this notion after the following example.

5.3 System Dynamics

We consider an elementary function in electrical and biological systems – the first-order low-pass filter. The electrical function equivalent is depicted in Figure 5.4.

The electrical network representation of Figure 5.4 is the most common functional model in electrical and biologic systems. Therefore, we will expand the fundamental equations necessary for understanding.

From elementary electrodynamics equations used for the capacitor:

$$Q = CU_C \tag{5.6}$$

This equation has been used in Chapter 3 of the modeling introduction. Here Q represents the amount of electrical charge of the capacitor C and U_C

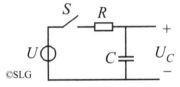

Figure 5.4 General electrical network representation of a low-pass resistor capacitor combination excited by means of a voltage step U and switch S.

represents the voltage across the capacitor as a result of the amount of charge and the capacitance value. Equation 5.6 represents a static state of charge Q in and voltage U_C across a capacitor. This situation becomes dynamic when we have changes of charge Q and U_C as a function of time.

When we encounter an amount of charge ΔQ transported in a defined amount of time Δt, we have the situation of a charge current,

$$I = \frac{\Delta Q}{\Delta t} \tag{5.7}$$

When we apply this dynamic operation to Equation (5.7) we get

$$I = \frac{\Delta Q}{\Delta t} = C\frac{\Delta U_C}{\Delta t} \tag{5.8}$$

Here the coefficient of capacitance C is time invariant and U_C is therefore a variable of time. Equations 5.6, 5.7, and 5.8 will generally be valid as long as the variations are linear functions of time. The moment we encounter a non-linearity, we have to apply the infinitesimal definition of local variations defined as

$$I = I_C = \frac{dQ}{dt} = C\frac{dU_C}{dt}, \tag{5.9}$$

when the infinitesimal limit value of $\Delta t => 0$.

The transfer function of the circuit of Figure 5.4 can be found as follows. The moment switch S is closed, and we assume that the starting voltage of $U_C = 0$, we find the following equations.

The voltage across R equals U. This implies that the starting current through R and C equals:

$$I_R = I_C = \frac{U}{R} \tag{5.10}$$

Furthermore, the following equation is valid

$$U = RI_C + U_C \tag{5.11}$$

With

$$I_C = C\frac{dU_C}{dt}, \tag{5.12}$$

we find the differential equation

$$U = RC\frac{dU_C}{dt} + U_C, \tag{5.13}$$

and after expansion we have

$$Udt = RCdU_C + U_Cdt \tag{5.14}$$

With $RC = \tau$ we find after integration the general solution

$$U_C = U\{1 - \exp(-t/\tau)\} \tag{5.15}$$

The behavior of U_C as a function of time is depicted in Figure 5.5.

Similarly, when the capacitor of Figure 5.4 is charged to voltage U, a discharge operation via a resistor results in a negative exponential decay. This situation is given in Figure 5.6. When switch S is closed in Figure 5.6, the charged capacitor C with voltage U_C will be discharged according to

$$U_C(t) = U\exp(-t/\tau) \tag{5.16}$$

This relationship is graphically depicted in Figure 5.7.

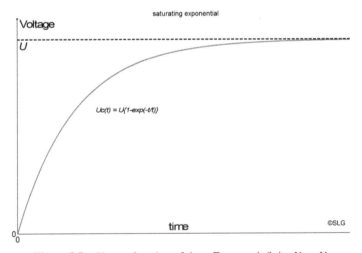

Figure 5.5 U_c as a function of time. For $t => $ infinite $U_c = U$.

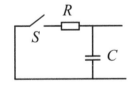

Figure 5.6   Discharge circuit.

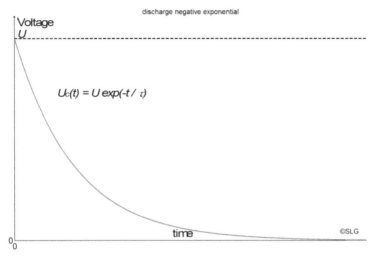

Figure 5.7 Discharge characteristic of U_C from Figure 5.6.

5.4 Frequency Response of the First-order Low-pass Section

According to Equation 5.12, we now can find the frequency response of a resistor capacitor network by means of the Laplace transform operation. This means that the equations with time as the variable are transformed to the $1/t$ or the frequency domain. In the case of the Laplace transform, we introduce the variable $s = j\omega = j.2\pi f$ representing the complex frequency. We can write Equation 5.12 as

$$\frac{dt}{C} = \frac{dU_C}{I_C},\tag{5.17}$$

representing the reactance of C as a frequency-dependent resistance.

$$\mathfrak{A}\left(\frac{dt}{C}\right) = \int_0^\infty \left(\frac{dt}{C}\right) \exp(-st),\tag{5.18}$$

then

$$\int_0^\infty \frac{1}{C}\exp(-st)dt = \frac{1}{C}\int_0^\infty \exp(-st)dt = \frac{1}{C}\exp(-st)]_0^\infty = \frac{1}{sC}\tag{5.19}$$

From Equation (5.19) we can appreciate that the term $\frac{1}{sC}$ will decrease with increasing values of s.

Now we can analyze the circuit configuration of Figure 5.8.

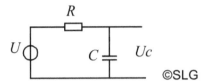

Figure 5.8   First-order passive low-pass filter.

The impedance Z_R of R equals the value of R as $Z_R = R$

The impedance Z_C of C equals $Z_C = \frac{1}{sC} = \frac{1}{j\omega C}$ with $\omega = 2\pi f$

The signal current through Z_R and Z_C can be written as

$$I = \frac{U}{Z_R + Z_C},\qquad(5.20)$$

resulting in an output voltage across C

$$U_C = Z_C I = \frac{Z_C U}{Z_R + Z_C} = \frac{U/sC}{R + 1/sC} = \frac{U}{sRC + 1}\qquad(5.21)$$

The voltage transfer function can now be expressed as

$$\frac{U_C}{U} = \frac{1}{sRC + 1}\qquad(5.22)$$

The product $RC = \tau$ represents the time constant of the low-pass network. When we look at the frequency behavior of the low-pass network, we expand the transfer function as follows.

with $s = j\omega$, $RC = \tau$ and $j^2 = -1$

$$\frac{U_C}{U} = \frac{1}{j\omega\tau + 1} = \frac{-j\omega\tau + 1}{(j\omega\tau + 1)(-j\omega\tau + 1)} = \frac{1 - j\omega\tau}{(\omega\tau)^2 + 1}\qquad(5.23)$$

Resulting in a real part of $\frac{U_C}{U} = \frac{1}{(\omega\tau)^2 + 1}$ and an imaginary part $\frac{U_C}{U} = \frac{-j\omega}{(\omega\tau)^2 + 1}$.

In the complex plane we depict this as shown in Figure 5.9.

The modulus of the transfer function

$$\left|\frac{U_C}{U}\right| = \left|\frac{1}{sRC + 1}\right| = \sqrt{\left(\frac{1}{(\omega\tau)^2 + 1}\right)^2 + \left(\frac{\omega\tau}{(\omega\tau)^2 + 1}\right)^2}\qquad(5.24)$$

$$= \sqrt{\frac{1}{\{(\omega\tau)^2 + 1\}^2} + \frac{\omega^2\tau^2}{\{(\omega\tau)^2 + 1\}^2}}$$

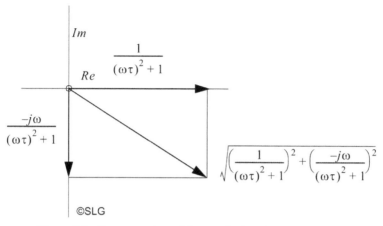

Figure 5.9   Representation of the transfer function of Figure 5.8.

A common representation of the frequency behavior of the transfer function $H = \frac{U_C}{U}$ is displayed in a so-called Bode diagram. Such a plot has a modulus and a phase diagram display of $H(j\omega)$ as depicted in Figure 5.10,

$$\text{The phase behavior of } \left| \frac{U_C}{U} \right| \text{ is defined as } arctg \left(\frac{Im(U_C/U)}{Re(U_C/U)} \right) \quad (5.25)$$

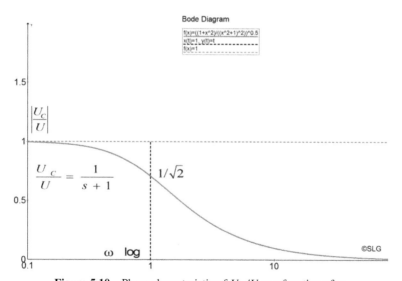

Figure 5.10   Phase characteristic of U_C/U as a function of ω.

With this example, we introduced a method to analyze the frequency and phase behavior of a system with the Laplace transform of first-order low-pass circuits. For a further study of this subject, the excellent work about feedback systems written by Astrom and Murray [6] is highly recommended.

5.5 Cascading System Blocks

An important property of cascaded stages is the amplification of transfer functions and signals of which the effects are shown in Figure 5.11.

From Figure 5.12, we derive that the transfer from output Y to input X equals

$$\frac{Y}{X} = G \qquad (5.26)$$

5.6 Example of a Feedback System

Let us go back to our amplifier block example of Figure 5.1. Furthermore, we introduce a new network element, the adding element, or comparator. The adding element is depicted in Figure 5.13.

If we combine the first adding element of Figure 5.13 with the amplifier from Figure 5.1, we have the simplest feedback loop configuration. The choice for the first adding component is based on the fact that we want to describe a feedback system with negative feedback in order to realize stabilization. The result is that the output signal (B) will be subtracted from the input signal (A).

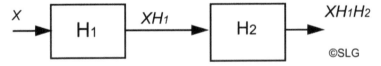

Figure 5.11   Cascade of two stages with transfer functions H_1 and H_2 with the multiplication effect.

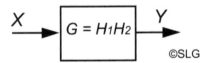

Figure 5.12   Representation of the combination of amplifier gain.

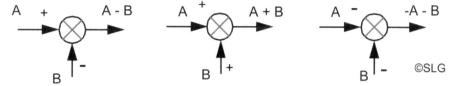

Figure 5.13   Three different adding elements.

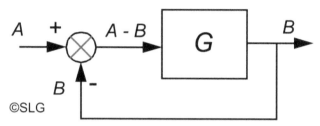

©SLG

Figure 5.14   Unity negative feedback loop.

We will use Figure 5.14 to analyze and calculate the closed-loop situation. This example is very important because it shows the analysis based on a recursivity that will be shown in the calculation. For clarity, we define the signal $A - B$ as the signal in the feed forward path. The *feed forward path* in a system is fundamentally to be distinguished from the notion of feed forward, which is a method to compensate for unwanted output deviations. Feed forward is thus a fundamentally different way to stabilize or compensate system behavior, compared to feedback.

The value A is called the input reference. The value B is the output signal of the closed-loop configuration. If we follow the calculation schematic from Figure 5.14, we get:

$B = G(A - B)$ then we have $B = GA - GB$ resulting in $B + GB = GA$ or $B(1 + G) = GA$ and then we can finally write

$$\frac{B}{A} = \frac{G}{G+1} \tag{5.27}$$

From Equation (5.27), we can easily read that for large values of G it follows that

$$\frac{B}{A} = 1 \tag{5.28}$$

The term negative feedback is here valid because the sign of the output signal B is negative in comparison to the reference signal A.

In the case that we would apply positive feedback, the output signal would be in phase with the input signal, so the system will directly cumulate in an infinite situation from which no input-to-output response can be expected anymore. The system has then been locked in a final state.

The positive feedback transfer function will result in

$$\frac{B}{A} = \frac{G}{G-1} \tag{5.29}$$

The phenomenon of instability can be illustrated with the following example of a linear system comprising different conditions of feedback. In Figure 5.15, the configuration of a unity feedback amplifier of Figure 5.14 is used to illustrate the effects of the phase of the input signal and output signals.

When we close the loop, we find a differential signal depicted as the black sinusoidal wave which is in opposite phase with the input signal. In this situation, the feedback result would be a sinusoidal signal in phase with

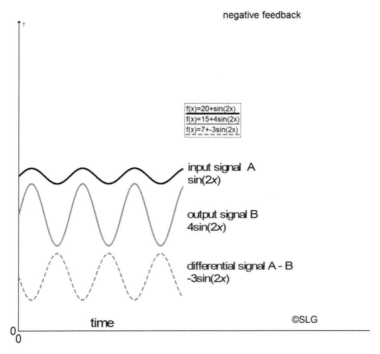

Figure 5.15   The input signal A is represented by the black sinus function and the open-loop output signal is represented by the gray sinus resulting in a stable feedback configuration.

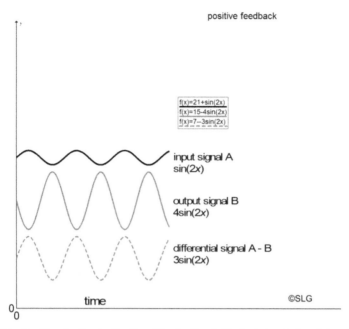

Figure 5.16 Positive feedback situation. The feedback (dashed sinus) signal A − B is now in phase with the input signal.

the input signal. We would encounter the following situation depicted in Figure 5.16.

According to the signal situation of Figure 5.16, the feedback signal is now in phase with the input signal resulting in a multiplicative escalation of the output signal when the loop gain $G > 1$. If $G < 1$, the output signal would always be smaller than the input signal preventing a multiplicative signal growth. In such a case, the amplifier configuration would have no realistic purpose because the output signal would return to zero!

For values of

$$G \approx 1, \qquad (5.30)$$

we see that

$$\frac{B}{A} => \infty \qquad (5.31)$$

i.e., instability. In general, this is the condition $G \approx 1$, where we define the system to be unstable as long as

$$G > 1 \qquad (5.32)$$

5.6.1 Feed Forward Compensation

Another kind of control mechanism to stabilize a system in a way we find suitable is known as feedforward and is commonly applied in situations where the properties of the system are well known and results in a compensation of unwanted transfer anomalies.

This system configuration is depicted in Figure 5.17.

In such a situation, we compensate for situations and system responses coming ahead based on the knowledge we have about the main system response.

A very striking well known example occurs when we drive a vehicle. Driving an automobile (G_1), the moment we see a curvature of the road well ahead, we would have automatically already taken actions (H_1) to compensate for the disturbing effect of the curve in the road in order to keep the vehicle on track (B).

In a dynamic situation, this kind of system control needs a separate feedback signaling path faster than the process that has to be controlled in order to maintain stable system control. This is also well demonstrated by the eye–hand feedback control where fast feedback (eye), with the speed of light, results in compensating actions on the hand movements. We can also recognize these properties in the glucose–insulin system where the slow metabolic process is regulated by the pancreas, etc., and the fast, looking ahead (feed forward action) is controlled by the central nervous system and the nerves. Although there are indications that the HPT control system has similar fast feedback mechanisms via the neural control on TRH signaling in the HP, the main control mechanism in the HPT axis is still based on negative

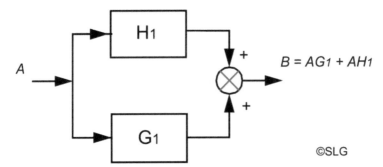

Figure 5.17   Output stabilization by feed forward compensation.

feedback effects of [FT4]. In a following chapter, we will discuss the HPT feedback loop in more detail.

5.7 Discussion

From the previous parts in this section about system theory, we discussed the open system constructed by a cascade of one or more amplifying sections in series. The open system is considered here as linear with the example of an audio amplifier without dynamics. Then we can examine the static transfer properties of such an open chain. The open system is quite comparable to the open loop of a person without an operating thyroid. The gain factor of such an open loop determines the steepness of the open-loop transfer. In the following Chapter 6, the steepness factor will be recognized as the exponential coefficient φ of the transfer characteristic of the HP unit in the brain. This steepness represents the sensitivity for input signal variations.

A high gain factor results in strong variations at the output with a standardized input of 1.

A low gain factor can result in a fivefold lower output response at the same input, representing a signal attenuation. Practical observations where hypothyroid patients were examined based on their titration of doses using levothyroxine (L-T4) resulted in various steepness factors of their HP. From the calculated values of φ, it became clear that persons with a high value of φ were more sensitive to variations in their daily dose of L-T4. Because the HP characteristic is non-linear, we can determine the local steepness by means of differentiation at a certain point on the HP curve in which φ is the determining factor. Normally the HPT system is presented as a closed loop in which the output is governed by the internal set point value for [FT4] and the loop gain. These issues will be further discussed in Chapter 9.

References

[1] Roelfsema, F., and Veldhuis, J. D. (2013). Thyrotropin secretion patterns in health and disease. *Endocr. Rev.* 34, 619–657. doi: 10.1210/er.2012-1076

[2] Goede, S. L., and Leow, M. K. (2013). General error analysis in the relationship between free thyroxine and thyrotropin and its clinical relevance. *Comput. Math. Methods Med.* 2013:831275. doi: 10.1155/2013/831275

[3] Goede, S. L., Leow, M. K., Smit, J. W. A., and Dietrich, J. W. (2014). A novel minimal mathematical model of the hypothalamus–pituitary–thyroid axis validated for individualized clinical applications. *Math. Biosci.* 249, 1–7. doi: 10.1016/j.mbs.2014.01.001

[4] Goede, S. L., Leow, M. K., Smit, J. W. A., Klein, H. H., and Dietrich, J.W. (2014). Hypothalamus-Pituitary-Thyroid Feedback Control: implications of mathematical modeling and consequences for thyrotropin (tsh) and free thyroxine (FT4) reference ranges. *Bull. Math. Biol.* 76, 1270–1287 doi: 10.1007/s11538-014-9955-5

[5] Leow, M. K., and Goede, S. L. (2014). The homeostatic set point of the hypothalamus-pituitary-thyroid axis - maximum curvature theory for personalized euthyroid targets. *Theor. Biol. Med. Model.* 11:35. doi: 10.1186/1742-4682-11-35

[6] Aström, K. J., and Murray, R. M. (2009). *Feed Back Systems*. Princeton, NJ: Princeton University Press.

6

The Mathematical Relationship between [FT4] and [TSH]

"The essence of mathematics is not to make simple things complicated, but to make complicated things simple."

–Stanley Gudder (1937–present)

6.1 Introduction

6.1.1 History

In the early days of thyroid research, there were questions about the factors influencing the operation of the thyroid. The earliest known reports are found in China about 1,600 years before our calendar. There is also a detailed drawing made in the 15th century of the anatomy around the area of the thyroid by Leonardo da Vinci.

The treatment of goiter was first provided by Prout in 1821 whereas in 1835, Robert Graves published his research on goiter treatment. Later on, Sir William Gull recognized among others the condition of hypothyroidism. In those days, there were no international standards to communicate published research findings because of language and geographical barriers. This led to much misunderstandings and double discoveries both in the United Kingdom and Europe. For instance, while autoimmune-mediated hyperthyroidism is named Graves' disease after its Irish discoverer (Robert James Graves) who described it in 1835, it is better known as Basedow's disease in Europe in honor of the German Karl Adolph von Basedow who independently reported the same condition in 1840.

A first report of a thyroidectomy in Germany by Ludwig Rehn was published in 1888. It was observed that persons with a thyroidectomy developed the same symptoms as persons with a known hypothyroidism. The recommended treatment for hypothyroidism was the use of a lightly fried sheep's thyroid taken once a week. The condition of struma lymphomatosoma was reported and published by Hashimoto Hakaru, now recognized as the most common cause of hypothyroidism and designated Hashimoto thyroiditis. The discovery that iodine plays a major role in thyroid physiology led to the isolation of thyroxine by Edward Calvin Kendall in 1914. Only in 1926, Harrington of the University Medical School in London elucidated the correct chemical structure of thyroxine, which led to the synthesis of what we now know as L-thyroxine or L-T4. Pitt-Rivers and Gross discovered and synthesized in 1952 the biologically more active tri-iodothyronine. In 1953, Werner published a paper about the pituitary–thyroid relationship in normal and disordered thyroid states [1]. Here the first steps were taken to find out how this mechanism could operate.

Toward a more concrete idea of the physiological mechanism, early work was published by Danziger and Elmergreen in 1956 [2]. In this paper, they described the secretory dynamics of the thyroid gland and the control mechanism with the introduction of non-linear differential equations. This is one of the first attempts to model the thyroid system in a quantified mathematical way. This work was followed by Roston in 1959 [3]. Here the conditions and modeling of homeostatic situations of the thyroid-pituitary system were analyzed with differential equations and solved by means of Laplace-transformed algebraic equations. Still, this work struggles with the dynamical aspects caused by a disturbance or change. Apparently, a common weakness among these earlier works is the sheer lack of the notion that a closed-loop control mechanism cannot be analyzed in this way. This obfuscated an improved understanding of the homeostatic equilibrium state of the hypothalamus–pituitary–thyroid (HPT) system. It is therefore our collective goal to illustrate how such a problem can be finally solved by our theoretical framework as discussed in this book.

Another approach to understand the thyroid hormone relationships popular with many researchers is to resort to the application of statistical methods to analyze massive amounts of thyroid function data. The Thyroid Manager, the freely available overview of clinical and physiological thyroid research, mentions the relationship between concentrations of free thyroxine and thyroid stimulating hormone (FT4 and TSH) in a typical thyroid function test (TFT) denoted respectively by [FT4] and [TSH] with a diagram. A recent study used TFT scatter plots published by Jonklaas et al. in 2008 [4]. In this study, a new detection method, tandem mass spectrometry, was used to identify individual sets of [FT4] and [TSH] and presented as a scatter plot. From their perspective, the researchers were identifying the log-linear relationship of [TSH]–[FT4] and explained that this relationship was best characterized by the use of [FT4] tandem-mass spectrometry. The common derived idea is presented in Figure 6.1. Here, the bivariate ranges of [FT4] are depicted horizontally while [TSH] values are depicted vertically in a two-dimensional picture. The measured TFT values were taken from healthy volunteers and each point represents the individual homeostatic point of the persons involved. It is important to appreciate that sudden perturbations to the euthyroid homeostatic equilibrium state in healthy people cause only a temporary turmoil because the homeostatic condition is quickly restored and the set point maintained by a robust feedback control. We define a euthyroid set point as the thyroid homeostatic hormone equilibrium of [FT4] related to the belonging value of [TSH].

From the scatter plot of Figure 6.1, we have an idea about the area and range in the [FT4]–[TSH] plane where the various homeostatic points of a sampled population are found. This is a good example of descriptive statistics that provides a rough indication of natural appearance for the values of the investigated subject. The only mutual relationship between the measured homeostatic points is the fact that their values emerge in a certain bounded area and that they were retrieved from humans in a certain population.

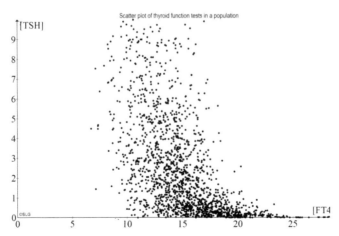

Figure 6.1 Scatter plot of euthyroid homeostatic points of a group of unrelated individuals with [TSH] and [FT4] represented using linear scales.

Furthermore, it is obvious when we appreciate the presentations in Figure 6.1 that the mutual distinction in the lower end of the [TSH] scale is rather blurred by the relative high density of TFT values.

In order to get a better view of this situation the presentation with a logarithmic scale for [TSH] solves this presentation issue and can be appreciated from Figure 6.2.

The log [TSH]–linear [FT4] presentation of Figure 6.2 is exemplary for all other scatter plot presentations in clinical publications about this subject. This is also the reason why the special area where the regression line of the scatter plots of [FT4] and [TSH] could demonstrate a distinct curvature, in the case of a linear [TSH]–linear [FT4] axis, has been missed by all researchers using the log-linear axes arrangement.

However, when we try to find a mutual relationship of [FT4] and [TSH] between these measured data points of totally unrelated individuals, the situation is radically changed. Inferential statistical techniques normally apply regression lines that are basically attempts to find a curve fit to combine all measured results in a single equation or "model." However, this approach makes no sense in elucidating an individual [FT4]–[TSH] relationship as depicted in Figure 6.2 because the [FT4] of one person can never have any

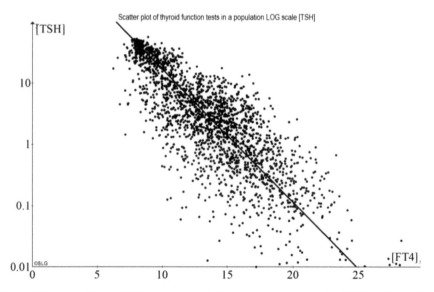

Figure 6.2   Logarithmic [TSH] scale presentation and a linear scale for [FT4] with a solid linear regression line.

effect at all on someone else's [TSH] and vice versa. This inferential kind of statistics is completely confused with the statistical methods applied in physics where the behavior of a population of identical particles with mutual interactions in a closed system is studied.

The word "model" has lost its meaning because there is no existing theory that describes and explains the influence between the measured results of independent objects. This is obvious because the homeostatic set point value of each individual in a scatter plot has no relationship to the set points of other individuals in the dataset [5]. The regression line, supposedly to represent a "model," implies directly that the set point of person A would have a direct "supernatural" influence on the set point of person B. Fortunately, this is not the case! How would person B feel when his [TSH] would drop because of an [FT4] rise in person A and vice versa? This is clearly nonsensical and therefore the set points of person A and person B are not related! This is a common pitfall in all kinds of clinical research on populations.

The misinterpretation in this sense is well documented in various publications which are representative of other similar clinical research results from the early 1960s to 2016 [5–13]. These examples of misinterpretation of statistical results are mentioned here to show how such a research strategy fails to elucidate physiological relationships and mechanisms where connections between biological variables are only meaningful when interrogated at the individual rather than population level.

6.2 Linear-logarithmic Relationship between [FT4] and [TSH]

Given all the failed mathematical modeling attempts to describe the dynamics of the HPT system and to find relationships between [FT4] and [TSH] [14–20], the scientific approach is to analyze the measured data individual by individual. In this treatise, the relationship between [FT4] and [TSH] here is defined as the collection of stable equilibrium states measured at a certain moment in a single person. As we will discuss later on in detail, the primary reaction of the HPT system is based on the local detection and conversion of [FT4] to intracellular T3. This results in a stable equilibrium state of concentrations FT4 which is supported by the relative fast reacting [TSH] equilibrium state because of the relative short half-life of 1 h.

As mentioned in Chapter 4, clinical results based on scatter plots were flawed as far as thyroid physiology is concerned. Theoretically, when the TFTs of healthy individuals are presented in large enough numbers, the result should be a perfectly circular-shaped data cloud in the log-linear axis presentation from which no regression line could be found in whatever direction because the cloud would be perfectly symmetrical around the vertical mean value of the [FT4] axis! [21] In a huge dataset belonging to different people with a heterogeneous mix of TFTs comprising both perfectly individualized euthyroid set point values and others whose TFTs are not perfectly euthyroid, we will see a pattern of asymmetry that could suggest a direction to the log-linear relationship resembling the actual log-linear relationship that really applies to that between [TSH] and [FT4] of individual persons such as what we see in almost all publications that analyzed population TFTs. This widespread misunderstanding endures for as long as inferential statistical methods are inappropriately used for modeling of physiological processes.

The identification of the logarithmic-linear relationship between [TSH] and [FT4] was earlier described in a clinical observation of Suzuki et al. [22]. Despite the erroneous statistical method and interpretation of the results, this relationship was correctly written by accident in a parameterized logarithmic form and resulted in straight lines when the relationship was plotted. They found:

$$\log[TSH] = b - a[FH] \tag{6.1}$$

This relationship places [FH], representing free thyroid hormone concentration, as we will call it later free T4 or [FT4], on the horizontal axis and [TSH] on the vertical logarithmic axis in a two-dimensional plane. They also stated that the equation should be simplified letting $a = 1$. Variations in a were not investigated. Later we will show that the value of a plays a crucial role in individual diagnostics and treatment. As an illustration, this relationship is plotted in Figure 6.3 with b as model parameter.

In Figure 6.3, we see that the [FH]–[TSH] relationship is depicted with a log scale for [TSH] and a linear scale for [FH]. Variation of model parameter, $b = \ln(2,500)$, $b = \ln(5,000)$, and $b = \ln(10,000)$ results in a horizontal shift of the function along the [FH] axis at a constant value for $a = 1$ where the angle determined by the value of a remains unchanged.

For comparison, it is illustrative to depict the results from Figure 6.3 in a linear scale for [TSH] values. This is shown in Figure 6.4.

The presentation in Figure 6.4 gives directly a fundamentally different view of the properties of the [FH]–[TSH] characteristics. The plot in Figure 6.5 shows the effects of a variation of model parameter a while parameter b is kept constant at $b = \ln(5,000)$.

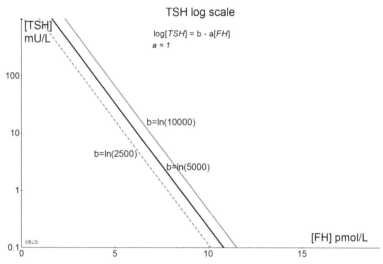

Figure 6.3 Graphical representation of Equation (6.1) where the [TSH] scale is logarithmic and the [FH] scale linear with $a = 1$ and with three different values for b.

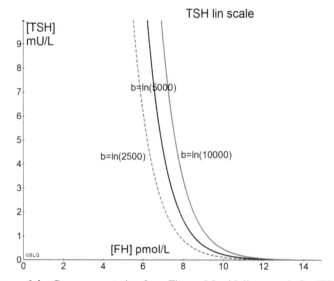

Figure 6.4 Curve presentation from Figure 6.3 with linear scale for [TSH].

Similarly, we show the results of Figure 6.5 with a linear scale for [TSH] in Figure 6.6.

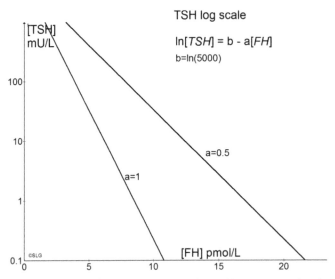

Figure 6.5   Graphical presentation of Equation (6.3) with constant value $b = \ln(5,000)$, $a = 1$ and $a = 0.5$.

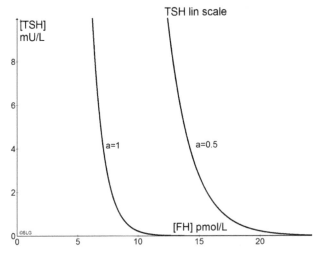

Figure 6.6   [FH]–[TSH] characteristics with linear [TSH] scale.

The postulated model of Suzuki et al. [22] was never investigated any further or tested on individual TFTs of hypothyroid patients who were treated with L-T4 to a euthyroid state. Unfortunately, this precluded the discovery of the properties and application of a parameterized exponential function describing the equilibrium characteristic of the [FT4]–[TSH] relationship because the vertical [TSH] axis was presented in a logarithmic scale! In the given presentation of Figures 6.3 and 6.5, we see straight lines from which no information can be obtained other than the position and the angle. This image changes totally when the [TSH] scale of these figures is presented as a linear one in Figures 6.4 and 6.6, respectively.

The rationale behind the log scale for [TSH] is that [TSH] encountered in health and disease ranges from 0.005 < [TSH] < 400 mU/L, a variation exceeding five orders of magnitude. Measured values of [FT4] on the other hand have a range spanning roughly 1 < [FT4] < 30 pmol/L normally, which is easily represented within the range of a linear scale and within an order of magnitude. A linear scale for [TSH] would have immediately revealed a prominent curvature in a certain area of the [FT4]-[TSH] plane. From the following, it will become clear that this curvature plays a central role in the location of the euthyroid set point.

The publication of an exponential relationship between [FT4] and [TSH] of Leow in 2007 [16], contained a crucial clinical data set that provided the first possibility to verify the current model.

The correct interpretation describing the relationship between [FT4] and [TSH] is a parameterizable exponential characteristic representing the collection of all possible equilibrium conditions of [FT4] and [TSH] in an individual, validated with clinical data and was published by Goede et al. (2014) [23]. The equation of Suzuki, Equation (6.1) can be transformed to a presentation in which the [TSH] axis is linear according:

$$\exp\{\log[TSH]\} = \exp\{b - a[FH]\} \tag{6.2}$$

$$[TSH] = \exp\{b - a[FH]\} \tag{6.3}$$

$$[TSH] = \exp(b)\exp\{-a[FH]\} \tag{6.4}$$

$$[TSH] = \frac{\exp(b)}{\exp\{a[FH]\}} \tag{6.5}$$

This log-linear relationship is actually only valid for equilibrium conditions in TFT during L-thyroxine dose titration from a hypothyroid situation to a euthyroid condition.

In a normal healthy person, the pair of [FT4]–[TSH] values represents the euthyroid set point which is relatively fixed at a stable value for [FT4] and a corresponding [TSH] that exhibits circadian variations [15]. This becomes apparent when a person is diagnosed with a primary hypothyroidism due to such etiologies as Hashimoto thyroiditis, post-radioiodine thyroid ablation, or total thyroidectomy. Because the hypothalamus–pituitary (HP) is no longer under the influence of the thyroid in such states, we encounter the situation of an open-loop system. The series of TFT values in such an open loop provides a very good picture of the possible curve or characteristic that can be found when the dots of the TFTs are connected. We have to keep in mind that even if the HPT loop is opened by a non-functional thyroid, the HP system still operates perfectly as an [FT4] sensor and [TSH] generator over a wide input $0.1 < [FT4] < 100$ pmol/L and output $0.01 < [TSH] < 1000$ mU/L range. When we examine the properties of the HP system, we find a complex control unit. Actually, the complexity is so enormous that we cannot imagine how many model parameters could be involved to present a complete analyzable model. It is however unnecessary to construct such a model that has no practical solution. Quite the opposite, it is usually much more insightful to derive a simple model that extracts the gist of the properties of the system with the minimum of variables analogous to the principle of Occam's razor.

6.3 Modeling the Curved [FT4]–[TSH] Characteristic

The following modeling exercise uses TFT data typical of many cases observed during thyroid hormone replacement in hypothyroid patients with negligible thyroid function where the TFTs were performed under well-controlled conditions. Let us assume that each patient has perfect adherence to daily L-thyroxine, that TFTs are measured at 6–8 weekly intervals, and blood samples are taken between 0700 h and 0900 h before breakfast and daily L-thyroxine so as to have a stable test condition [23]. Based on these conditions, we find the following situation as depicted in Figure 6.7.

With the help of a graphical mathematical tool (Graph 4.4) freely downloadable from the Internet [24], we can derive some insight into the [TSH]–[FT4] relationship.

Hypothetically, a novice without any knowledge of biology and thyroid physiology could possibly interpret this shape as the part of a quadratic polynomial. With one of the predefined fitting functions of Graph 4.4, we can find out if this could be true. In the Figures 6.8–6.11 we depict a series of fitting possibilities.

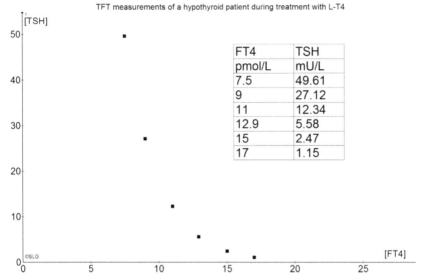

Figure 6.7 The plotted measurement results of a hypothyroid patient.

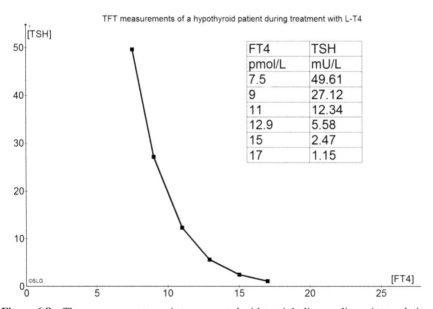

Figure 6.8 The measurements are interconnected with straight lines or linear interpolation.

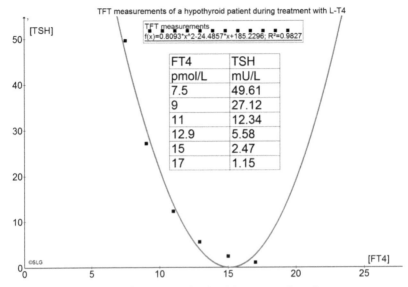

Figure 6.9   Polynomial fit of the measured results.

$y = 0.893x^2 - 24.4857x + 185.2296$ with a fitting quality of $R^2 = 98\%$ appears superficially quite good, though this would be rather implausible because it would be nearly impossible to find coordinates of ([FT4], [TSH]) that occur on the rising limb of this parabolic curve.

A second experiment is performed with a power fitting function.

The power function that is found as a fit $y = 657630x^{-4.6181}$ with a fitting quality of $R^2 = 94\%$ seems not bad either.

The third experiment is a fitting attempt with an exponential function.

$y = (989.63)(0.6706)^x$ with a fitting quality of $R^2 = 100\%$. We can rewrite y as $y = 989.63 \exp(-0.3995x)$.

We have in fact found and validated an exponential relationship between [FT4] and [TSH] which can be written in a parameterized form as $y = S \exp(-\varphi x)$ with $y = $ [TSH] and $x = $ [FT4] as a generalized expression. Thus,

$$[TSH] = S \exp(-\varphi[FT4]), \tag{6.6}$$

which is identical as the expression we found in Equation (6.5) where

$$\exp(b) = S, a = \varphi \text{ and } [FH] = [FT4] \tag{6.7}$$

This relationship is verifiable by clinical data which supported the exponential model with individualized parameters S and φ [23].

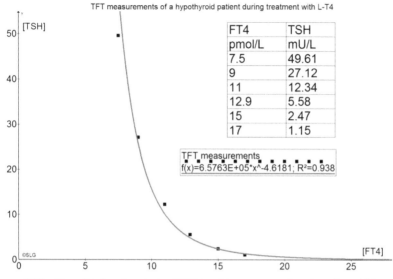

TFT measurements of a hypothyroid patient during treatment with L-T4

FT4	TSH
pmol/L	mU/L
7.5	49.61
9	27.12
11	12.34
12.9	5.58
15	2.47
17	1.15

TFT measurements
f(x)=6.5763E+05*x^-4.6181; R²=0.938

Figure 6.10 Thyroid function tests (TFTs) with a power function fit, which seems rather good.

6.4 Properties of the Exponential Function

Substituting y = [TSH] and x = [FT4], we can write Equation (6.6) as

$$y = S \exp(-\varphi x) \tag{6.8}$$

The exponential function of Equation (6.8) has the following properties:

1. This two-dimensional function is generally completely determined by only two coordinates (x_1, y_1) and (x_2, y_2).
2. The curvature K in a point of a function is defined as

$$f_K = K = \left(d^2y/dx^2\right)\left(1 + \left(dy/dx\right)^2\right)^{-3/2} = y''\{1 + (y')^2\}^{-1.5} \tag{6.9}$$

In a later section, we will discuss the derivation and the properties of the unique point where this maximum curvature is situated. In the following, the implications and consequences of various choices for all parameters involved will be discussed.

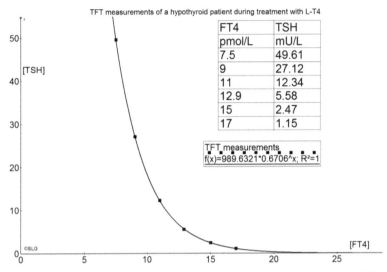

Figure 6.11   Thyroid function test measurements with a perfect exponential fit.

6.4.1 Effect of the Multiplier Model Parameter S

When the multiplier parameter of Equation (6.6), S, is varied over a certain range, the position of the function will translate along the [FT4] axis while the curve preserves the same shape at a constant value of φ. This shifting effect is depicted in Figure 6.12.

The factor S, a linear component of the thyrotropic system, is related to the [FT4] range and also (as we will discuss later) to the parameter determining the inhibition of [TRH]. Variation of S, with a fixed value for φ, horizontally translates the HP characteristic curve along the [FT4] axis as shown in Figure 6.12. When φ is fixed, we can appreciate from the first derivative of Equation (6.6) with respect to [FT4]:

$$\frac{d[TSH]}{d[FT4]} = -\varphi S \exp(-\varphi[FT4]) = -\varphi[TSH] \tag{6.10}$$

Notably, this derivative is the same in all points with the same value for [TSH] because the values of [TSH] and φ remain constant and fixed. This implies that the first derivative will not change when S is varied from

$$0 < S < \infty \tag{6.11}$$

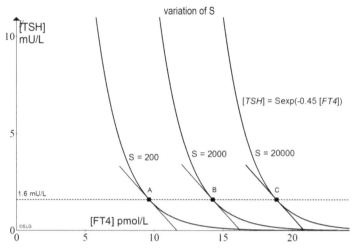

Figure 6.12   Horizontal translation of the exponential curve along the [FT4] axis by variation of S at a fixed value of φ resulting in the same value of the first derivative in points A, B and C.

6.4.2 Effect of the Exponential Model Parameter φ

The second model parameter φ represents the exponential factor. Variation of φ rotates the shape of the HP characteristic centered on a chosen set of coordinates P. Notably, φ and S are inter-related according to any [FT4]–[TSH] coordinate on the HP characteristic:

$$\varphi = \left(\frac{1}{[FT4]}\right) \ln\left(\frac{S}{[TSH]}\right) \tag{6.12}$$

In Figure 6.13, the folding effect of the variation of φ is shown, while the HP curves "rotate" around a defined point P.

The Figures 6.12 and 6.13 depict the theoretical range of values that φ and S can possibly assume under most clinical circumstances. Additionally, φ and S always form a parameter set describing a specific curve for a specific person. From Equation (6.10) it can be readily appreciated that the first derivatives at the intersecting point, P, are dependent only on the value of φ.

Calculation of the HP characteristic from two or more distinct [FT4]–[TSH] measurements of a single patient.

When two (or more) measured points $[FT4]_1$, $[TSH]_1$ and $[FT4]_2$, $[TSH]_2$ from an individual are distinctly separated, measured as TFT during the

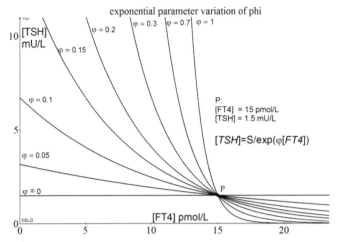

Figure 6.13   Different HP curves intersecting at [FT4] = 15 pmol/L and [TSH] = 1.6 mU/L. Here the array of curves is shown with theoretical values of $0 < \varphi < 1$. In practice, φ can assume values up to 1.5.

process of treatment, the HP characteristic can be plotted based on calculated parameters S and φ:

Using the expression and substitutions of Equation (6.6), we find
$y = [TSH]$ and $x = [FT4]$, we can write Equation (6.6) as $y = S \exp(-\varphi x)$
The measured coordinates of the points P_1 ([FT4]$_1$, [TSH]$_1$) and P_2 ([FT4]$_2$, [TSH]$_2$) are then expressed as:
$P_1\ (x_1, y_1)$ and $P_2\ (x_2, y_2)$ we can write

$$y_1 = S \exp(-\varphi x_1), \tag{6.13}$$

and

$$y_2 = S \exp(-\varphi x_2), \tag{6.14}$$

dividing Equation (6.13) by Equation (6.14) results in

$$\frac{y_1}{y_2} = \frac{S \exp(-\varphi x_1)}{S \exp(-\varphi x_2)} = \exp(-\varphi x_1) \exp(\varphi x_2) = \exp\{\varphi(x_2 - x_1)\}, \tag{6.15}$$

resulting in

$$\varphi = \left(\frac{1}{([FT4]_2 - [FT4]_1)} \right) \ln \left(\frac{[TSH]_1}{[TSH]_2} \right) \tag{6.16}$$

$$S = [TSH]_1 \exp(\varphi [FT4]_1) = [TSH]_2 \exp(\varphi [FT4]_2) \tag{6.17}$$

With the availability of additional points, i.e., ([FT4]$_3$, [TSH]$_3$), ([FT4]$_n$, [TSH]$_n$), etc., it is possible to verify the value of the parameters, S and φ, just by iterating the same calculation procedure between the third point and the other two points. The validity of measured TFT values is evaluated by a separate model selection procedure described in the dissertation of Martin Middelhoek [25]. Model selection and parameter identification provide a filtered result that excludes TFTs but are not part of the selected model space and are identified as outliers.

6.5 Dynamic Signal Transfer of the HP Characteristic

The HP characteristic is a collection of all possible equilibrium positions of [FT4] and [TSH] in an individual. The euthyroid homeostatic equilibrium is the most common situation in healthy people. However, under pathological conditions like Graves' disease, the thyroid has lost control over thyroid hormone synthesis because autoantibodies mimicking the TSH molecules (TRAb) activate the thyroid to produce T4 and T3 without the control of the normal feedback loop. This results in pathologically elevated [FT4] levels which in turn suppresses [TSH] according to the HP characteristic to levels as low as those beyond the detection limits of the TSH immunoassays. Nevertheless, these deviations on the HP curve away from the euthyroid area remain still points of equilibrium for a given individual even if they belong to a pathological state. A similar situation occurs when the thyroid fails as a consequence of thyroidectomy or other causes that result in low [FT4]. Hence, these points of equilibrium can be either pathological or normal (i.e., euthyroid). With these notions, we can discuss the properties of the dynamic behavior of hormone signaling. In Figure 6.14, we depict an example of the HP characteristic.

When input variations around a certain point of operation are introduced, as can be appreciated from Figure 6.14 where sinusoidal variations occur around the point [FT4] = 15 pmol/L,

$$f_{[FT4]} = 15 + 0.5\sin(\omega[FT4]), \tag{6.18}$$

the output variations

$$f_{[TSH]} = 1.6 - 0.35\sin(\omega[TSH]), \tag{6.19}$$

can be found as a transfer of the input signal on a small linear part on the HP characteristic indicated as a dashed tangent in P. The dynamic transfer function is then found as the first derivative of the HP characteristic

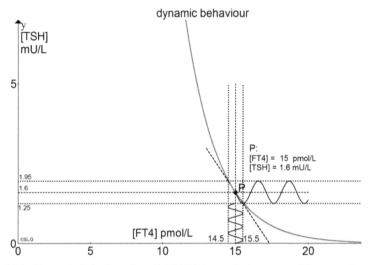

Figure 6.14   Dynamic behavior of small [FT4] input variations around the operating point P.

Equation (6.6) and results in the function of Equation (6.20). We define this transfer function as:

$$G_{HP} = \frac{d[TSH]}{d[FT4]} = -\varphi[TSH] \qquad (6.20)$$

6.6 Generalized Expression of the HP Characteristic

Until now, we only investigated the primary effect of [FT4] as main variable in the expression of Equation (6.6). However, there can also be a measurable influence of excessive externally administered [FT3] as a result of combination T4/T3 treatment for hypothyroidism [26]. Exogenous T3 administration superimposes with ambient [FT3] to augment the intracellular T3 (from deiodinated [FT4]) inhibitory signal on TRH and TSH of the hypothalamus and pituitary, respectively. This results in a significant drop of [FT4] at the set point of the HP curve which becomes shifted to the lower end of the [FT4] axis while the set point value of [TSH] is maintained. In effect, S will thus assume a lower value and the value of φ remains the same in this situation.

The generalized expression for the HP characteristic can then be written as

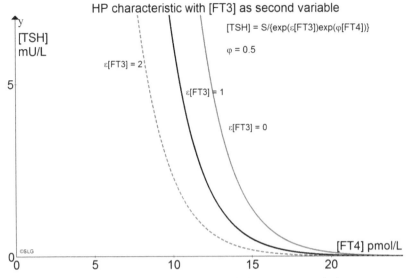

Figure 6.15 HP characteristic with [FT3] parameter variations.

$$[TSH] = \frac{S}{\exp(\varepsilon[FT3])\exp(\varphi[FT4])} \qquad (6.21)$$

This expression results in a reduction of the effect of model parameter S because of the increase of $\exp(\varepsilon[FT3])$ with increasing [FT3].

In Figure 6.15, we depict the effects of increasing inhibitory effect on [TSH] of externally administered T3 in the position of the characteristic.

6.7 Discussion

In this chapter, we discussed the quest to elucidate a correct and validated relationship between [FT4] and [TSH]. The parameters of this relationship are found with the individual TFT values of hypothyroid persons during their titration period from hypothyroidism to euthyroidism with L-T4 medication. The success of the model is impressive as the data fit is often better than 95% and represents a monumental milestone of the foundation to explore the HPT system and the related unique euthyroid set point position of any given individual. Based on the set point theory, we also find the conditions for a normal operating thyroid in a closed-loop situation. These findings will be discussed in the following chapters.

References

[1] Werner, S. C. (1953). Pituitary thyroid relationship in normal and disordered thyroid states. *Bull. N. Y. Acad. Med.* 29, 523–536.

[2] Danziger, L., Elmergreen, G. L. (1956). The thyroid-pituitary homeo-static mechanism. *Bull. Math. Biophys.* 18, 1–13.

[3] Roston, S. (1959). Mathematical presentation of some endocrinological systems. *Bull. Math. Biophys.* 21, 271–282.

[4] Jonklaas, J., and Soldin, S. J. (2008). Tandem mass spectrometry as a novel tool for elucidating pituitary-thyroid relationships. *Thyroid* 18, 1303–1311. doi: 10.1089/thy.2008.0155

[5] Fliers, E., Alkemade, A., Wiersinga, W. M., and Swaab, D. F. (2006). Hypothalamic thyroid hormone feedback in health and disease. *Prog. Brain Res.* 153, 189–207.

[6] van Deventer, H. E., Mendu, D. R., Remaley, A. T., and Soldin, S. J. (2008). Inverse log-linear relationship between thyroid-stimulating hormone and free thyroxine measured by direct analog immunoassay and tandem mass spectrometry. *Endocrinol. Metab. Clin. Chem.* 57, 122–127.

[7] Ercan-Fang, S., Schwartz, H. L., Mariash, C. N., and Oppenheimer, J. H. (2000). Quantitative assessment of pituitary resistance to thyroid hormone from plots of the logarithm of thyrotropin versus serum free thyroxine index. *J. Clin. Endocrinol. Metab.* 85, 2299–2303.

[8] Benhadi, N., Fliers, E., Visser, T. J., Reitsma, J. B., and Wiersinga, W. M. (2010). Pilot study on the assessment of the setpoint of the hypothalamus–pituitary–thyroid axis in healthy volunteers. *Eur. J. Endocrinol.* 162, 323–329. doi: 10.1530/EJE-09-0655

[9] Hoermann, R., Eckl, W., Hoermann, C., and Larisch, R. (2010). Complex relationship between free thyroxine and TSH in the regulation of thyroid function. *Eur. J. Endocrinol.* 162, 1123–1129. doi: 10.1530/EJE-10-0106

[10] Fitzgerald, S. P., and Bean, N. G. (2016). The relationship between population T4/TSH set point data and T4/TSH physiology. *J. Thyroid Res.* 2017:6917841. doi: 10.1155/2016/6351473

[11] Hadlow, N. C., Rothacker, K. M., Wardrop, R., Brown, S. J., Lim, E. M., and Walsh, J. P. (2013). The relationship between TSH and free T4 in a large population is complex, non-linear and differs by age and gender. *J. Clin. Endocrinol. Metab.* 98, 2936–2943.

[12] Hoermann, R., Midgley, J. E. M., Giacobino, A., Eckl, W. A., Wahl, H. G., Dietrich, J. W., et al. (2014). Homeostatic equilibria between free thyroid hormones and pituitary thyrotropin are modulated by various influences including age, body mass index and treatment. *Clin. Endocrinol.* 81, 907–915. doi: 10.1111/cen.12527

[13] Rothacker, K. M., Brown, S. J., Hadlow, N. C., Wardrop, R., and Walsh, J. P. (2016). Reconciling the log-linear and non-log-linear nature of the TSH-free T4 relationship: intra-individual analysis of a large population. *J. Clin. Endocrinol. Metab.* 101, 1151–1158. doi: 10.1210/jc.2015-4011

[14] Distefano, J. J., and Stear, E. B. (1968). Neuroendocrine control of thyroid secretion in living systems: a feedback control system model. *Bull. Math. Biophys.* 30, 3–26.

[15] Roelfsema, F., Pereira, A. M., Veldhuis, J. D., Adriaanse, R., Endert, E., Fliers, E., et al. (2009). Thyrotropin secretion profiles are not different in men and women. *J. Clin. Endocrinol. Metab.* 94, 3964–3967.

[16] Leow, M. K. (2007). A mathematical model of pituitary–thyroid interaction to provide an insight into the nature of the thyrotropin–thyroid hormone relationship, 2007. *J. Theor. Biol.* 248, 275–287.

[17] Eisenberg, M., Samuels, M., and DiStefano, J. J. (2008). Extensions, validation, and clinical applications of a feedback control system simulator of the hypothalamo-pituitary-thyroid axis. *Thyroid* 18, 1071–1085. doi: 10.1089=thy.2007.0388

[18] Eisenberg, M. C., Santini, F., Marsili, A., Pinchera, A., and DiStefano, J. J. (2010). TSH regulation dynamics in central and extreme primary hypothyroidism. *Thyroid* 20, 1215–1228. doi: 10.1089/thy.2009.0349

[19] Seker, O. (2012). *Modeling the Dynamics of Thyroid Hormones and Related Disorders, 2012, Submitted to the Institute for Graduate Studies in Science and Engineering in Partial Fulfillment of the Requirements.* Master thesis, Boğaziçi University, Beşiktaş.

[20] Pandiyan, B., Merrill, S. J., and Benvenga, S. (2013). A patient-specific model of the negative-feedback control of the hypothalamus–pituitary–thyroid (HPT) axis in autoimmune (Hashimoto's) thyroiditis. *Math. Med. Biol.* 31, 226–258. doi: 10.1093/imammb/dqt005

[21] Liu, J., Tang, Z., Zhang, J., Chen, Q., Xu, P., and Liu, W. (2016). Visual perception-based statistical modeling of complex grain image for product quality monitoring and supervision on assembly production line. *PLOS ONE* 11:e0146484. doi: 10.1371/journal.pone.0146484

[22] Suzuki, S., Nishio, S., Takeda, T., and Komatsu, M. Gender-specific regulation of response to thyroid hormone in aging. *Thyroid Res.* 5:1.

[23] Goede, S. L., Leow, M. K., Smit, J. W. A., Dietrich, J. W. (2014). A novel minimal mathematical model of the hypothalamus–pituitary–thyroid axis validated for individualized clinical applications. *Math. Biosci.* 249, 1–7. doi: 10.1016/j.mbs.2014.01.001

[24] Johansen, J. (2013). *Graph 4.4.2 A Graphical Mathematical Tool.* Available at: http://www.padowan.dk/graph/

[25] Middelhoek, M. G. (1992). *The Identification of Analytical Device Models.* Ph.D. thesis, Delft University Press, Delft.

[26] Jonklaas, J., Burman, K. D., Wang, H., and Keith, R., and Latham, K. R. (2015). Single-dose T3 administration: kinetics and effects on biochemical and physiological parameters. *Ther. Drug Monit.* 37, 110–118.

7

The Thyroid Gland and the Relationship Between [TSH] and [FT4]

"In mathematics we find the primitive source of rationality; and to mathematics must the biologists resort for means to carry out their researches."

–Auguste Comte (1798–1857)

7.1 Introduction

The thyroid gland physiology has been discussed in Chapter 2.

Earlier mathematical thyroid models were focused on the dynamic behavior, the size of the gland, and the synthesis processes. This is understandable because the changes and variations seem to be a variable that can be measured as a function of time and volume.

However, the basic condition for a full functionality is embedded in the physiological quality of the thyroid gland as a whole. Therefore, volume is a possible condition for secretory capacity, but in itself this value is meaningless. Other conditions like the availability of iodide, etc., are of primary concern.

Furthermore, the thyroid gland has possibly a secretion behavior which can be interpreted as a pulsatile current of T4 and T3 in a specific relation [1].

This pulsatility is directly related to the small oscillations resulting from a normal feedback control situation of the hypothalamus—pituitary—thyroid (HPT) system where the healthy thyroid is necessarily interacting with the hypothalamus – pituitary system. Pulsatility will not be observed in an open system where the remaining thyroid activity is overruled by a daily compensatory dose of L-T4.

However, these modeling attempts for analysis can never lead to a complete understanding of the relationships that are observed in the state of homeostasis. During the time-dependent signal processes most variations cannot be observed and most process parameters can never be found for a complete dynamic description. A literature overview from 1953 until the first workable model description published in 2014 by Goede et al. [2, 3] showed the deficiency and limitations of dynamical modeling of these systems. The complexity of biological systems is the result of millions of years of evolutionary development. Keeping this notion in mind, we can understand that most biological manifestations cannot simply be reduced to our library of understandings. However, we can try to use our imagination and common sense and try to find comparable and recognizable mechanical or electrical functions to see what really happens when we investigate a biological entity. It is not the internal organization, but moreover the final external manifestation of what we encounter and what we can measure in the human physiology as the final result. The most simple example is the fact that we eat, because we need food, we process the food internally (a really complicated process), we extract the nutritional components from the food, and excrete the non-usable material. In this simple sentence the complete intake, storage, digestion, and utilization of food is described. A similar observation and reduction to recognizable and mathematical formalization is the way to understand the investigated HPT phenomenon.

Because the dynamics of a process makes it practically impossible to come to a usable kind of model, the observation and measurement of an equilibrium opens the way to discover the underlying dynamic behavior.

The following modeling description is related to steady-state hormone currents and concentrations. Basically, the thyroid gland is a [TSH]-controlled [T4] and [T3] current source of which the transfer characteristic will be discussed. The notion of a current means that we define the transport of a certain amount of material per unit of time. It is a simple transformation of a time-dependent process to a stationary unit like current and finally the belonging equilibrium concentration. This allows the use of relatively transparent and simple mathematical descriptions.

7.2 Model of the Thyroid Gland

In general, the physiological gland mechanism can be regarded as a controlled generator of the specific substance they secrete. The thyroid gland is a manufacturing facility for the production and distribution of thyroid hormones, mainly known as thyroxine T4 and triiodothyronine T3.

The synthesis process will be distinguished from the distribution process. T4 and T3 are distributed in the ratio 8:1 and are regarded as currents or flows measured in a defined concentration as [T4] and [T3].

T4 secretion rate = 100 μg/24 h

The thyroid is therefore a T4 and T3 current generator the output of which is controlled by the amount of detected [TSH].

In most biological modeling descriptions it is common to use the Michaelis–Menten (MM) expression from enzyme kinetics to describe a gland secretory behavior [4].

A publication in 1988 and 1996 [5, 6] showed that the MM model expression has a limited validity range, and after a validation the saturating exponential approach proved to be superior to the flawed MM approximation as the best representation of the thyroid gland physiology. The secretory characteristic, where explicitly the unbound amount of T4 is noted as [FT4] of the thyroid, is expressed in the following equation.

$$[FT4] = A\{1 - \exp(-\alpha[TSH])\} \tag{7.1}$$

This equation stands for the collection of equilibrium points acquired from a steady-state [TSH] input current.

We can therefore regard Equation (7.1) as a quasi static characteristic that is valid for the individual in which it operates. The dynamic aspects can then be translated to time-dependent variations or signals in the case of a transient [TSH] or a transient in the output [T4] and [T3]. This has been discussed in Chapter 5, "Introduction in Systems Theory." A detailed transient behavior will be discussed in Chapter 14.

From Equation (7.1) the model parameter "A" represents the secretory maximum and "α" determines the steepness of the thyroid characteristic to reach the saturation level "A." In Figure 7.1 some variations of the exponential characteristic are presented with different model parameters.

From Figure 7.1 we can see the static characteristics of the thyroid output [FT4] related to the input signal [TSH]. These characteristics represent the so-called large signal model.

The dynamic properties are found when the input signal variations around a certain operating point result in related output variations of [FT4].

The changes in [TSH] variation can also be described with small concentration steps, after which we wait long enough to measure the equilibrium output result.

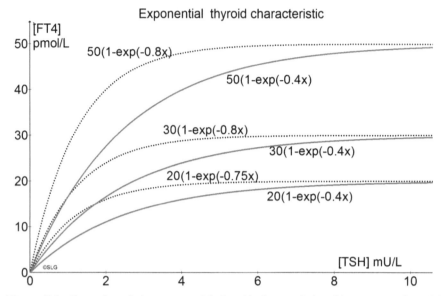

Figure 7.1   Examples of the exponential thyroid characteristic with a range of model parameters.

From this observation we can infer at least one time constant limiting the frequency response at a certain value. Other important dynamic parameters are the distribution behavior of [FT4] and [FT3], which will be discussed in detail in Chapter 14 about the "Appearance, Distribution and Half-life Behavior."

From Figure 7.2 we appreciate the point of operation P on the thyroid characteristic.

Variation of the input signal [TSH] around [TSH] = 1.5 mU/L with deviations of ±0.2 mU/L results in variations of [FT4] of ±1.25 pmol/L around 13.5 pmol/L.

The transfer factor G_T in point P is determined as

$$G_T = \frac{d[FT4]}{d[TSH]},\tag{7.2}$$

resulting in

$$G_T = \alpha A \exp(-\alpha[TSH])\tag{7.3}$$

From the mathematical form of Equation (7.3) we appreciate a behavior similar to the hypothalamus–pituitary (HP) characteristic where αA replaces S, α replaces φ.

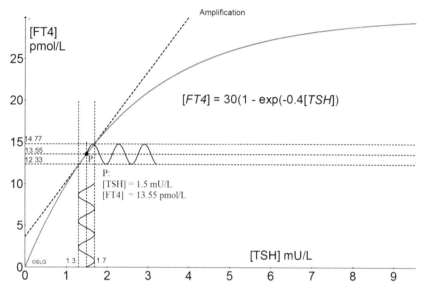

Figure 7.2   Sinus input signal variations around point of operation P on the [TSH] axis and output variations in [FT4] resulting in a signal transfer when relatively low frequency (0.01 Hz) varying changes in the concentrations are considered.

When in a healthy person the thyroid characteristic intersects with the hypothalamus – pituitary characteristic, we find the euthyroid homeostatic set point which indicates the level of operation for [FT4] in that individual.

The homeostatic set point is thus the common point of the HP curve as well as the thyroid characteristic.

Because the set point is derived from the properties of the HP characteristic, this property determines therefore the thyroid model parameters A and α. This relationship will be discussed in Chapter 8 about the "Point of Maximum Curvature [7]."

7.3 Discussion

The thyroid characteristic has the general property of saturation to a maximum value.

This is a natural phenomenon for all secreting glands in biological systems.

Every gland has a limitation with regard to the maximum secretory capacity because of the limited volume.

Based on findings described in the research of Keller et al. [5, 6] we choose the saturating exponential function as a model for the thyroid characteristic.

Later validations of the loop gain G_L in relation to the empirically established clinical reference ranges for [TSH] will demonstrate the correctness of this model.

References

[1] Mortoglou, A., and Candiloros, H. (2004). The serum triiodothyronine to thyroxine (T3/T4) ratio in various thyroid disorders and after Levothyroxine replacement therapy. *Hormones* 3, 120–126.

[2] Goede, S. L., Leow, M. K., Smit, J. W. A., Klein, H. H., and Dietrich, J. W. (2014). Hypothalamus-Pituitary-Thyroid feedback control: implications of mathematical modeling and consequences for thyrotropin (TSH) and free thyroxine (FT4) reference ranges. *Bull. Math. Biol.* doi: 10.1007/s11538-014-9955-5

[3] Goede, S. L., Leow, M. K., Smit, J. W. A., Dietrich, and J. W. (2014). A novel minimal mathematical model of the hypothalamus-pituitary-thyroid axis validated for individualized clinical applications. *Math. Biosci.* 249, 1–7. doi: 10.1016/j.mbs.2014.01.001

[4] Dietrich, J. W. (2002). *Der hypophysen-schilddrüsen-regelkreis. Entwicklung und endung eines nichtlinearen, Modells.* Berlin: Logos-Verlag.

[5] Keller, F., Emde, C., and Schwarz, A. (1988). Exponential function for calculating saturable enzyme kinetics. *Clin. Chem.* 34, 2486–2489.

[6] Keller, F., and Zellner, D. (1996). The 1-exp function as an alternative model of non-linear saturable kinetics. *Eur. J. Clin. Chem. Clin. Biochem.* 34, 265–271.

[7] Leow, M. K., and Goede, S. L. (2014). The homeostatic set point of the hypothalamus-pituitary-thyroid axis – maximum curvature theory for personalized euthyroid targets. *Theor. Biol. Med. Model.* 11:35. doi: 10.1186/1742-4682-11-35

8

The Hypothalamus–Pituitary (HPT) Set Point Theory

"Arc, amplitude, and curvature sustain a similar relation to each other as time, motion, and velocity, or as volume, mass, and density."

–Carl Friedrich Gauss (1777–1855)

8.1 Introduction

It is becoming an increasingly common scenario for hypothyroid patients on life-long thyroid hormone replacement (typically L-thyroxine or L-T4) to express symptomatic complaints despite being "biochemically euthyroid."

This state is defined by thyroid function tests (TFTs) comprising [FT4] and [TSH] and found within their respective reference ranges [1, 2]. In individual cases, discomfort may manifest as symptoms consistent with thyroid hormone excess or deficiency. This strongly suggests that the "normal range" values they have attained must somehow be either above or below what would have been the optimally paired [FT4]–[TSH] values of these "euthyroid" patients. Therefore, we have to consider that it is most likely that the [FT4]–[TSH] coupled optimum values are strictly individualized and unique for every given person. When we look at the various hypothalamus–pituitary (HP) characteristics discussed in Chapter 6 with linear scales for [FT4] and [TSH], we observe that the normal levels of [FT4] and [TSH] fall within the region where the negative exponential function has the greatest curvature, a curve segment on the function resembling the "knee" or "elbow" as it were. Based on this observation, we can investigate the properties of the HP function in more detail. From standard calculus, we know the algorithm to calculate the curvature of a two-dimensional function and to determine the point where the HP curvature represents a maximum. This is a unique point on the exponential function where changes in [FT4] have the most profound effects on variations in [TSH] from the point of maximum curvature to the lower values of [FT4]. Vice versa we find similar strong variations in [FT4] at relatively small variations in [TSH] just below the [TSH] value of the coordinate for the point of strongest curvature.

Given this interesting property, it is not surprising that the point of maximum curvature of a function would be strategically chosen by nature as an efficient method for optimal homeostatic control purposes in biological systems.

8.2 The Hypothalamus–Pituitary (HP) Control System

In healthy people, the HP unit is a sensitive metabolic sensor and master controller in a closed-loop thyroid hormone control system. As such, as is known from systems and control theory [3], a closed-loop system can be analyzed in the frequency and the time domains with a step and/or impulse response method. If these test methods cannot be applied, the alternative is to investigate the properties of the system in an open-loop mode. When the feedback loop is opened, as we have seen in patients where the thyroid hormone secretory function has failed, we can analyze the dose response properties of the HP in an interrogative and proportional way. This forms the basis for our validated parameterized negative exponential expression as was discussed in Chapter 6.

The closed negative feedback loop of the HP and thyroid needs a well-defined point of reference on which the narrow [FT4] range will be established and maintained. The intrinsic properties of the HP characteristic can be demonstrated to actually encode this reference value and hence provide a solution to find this unique point.

Although 3,5,3′-triiodothyronine, T3, is the key active thyroid hormone of the body, the response characteristics' analysis is confined purely to the relationship between [FT4] and [TSH]. Having alluded to this, it is crucial to appreciate the fact that most clinical data of hypothyroid patients treated with combination therapy using both L-T4 and L-T3 showed relatively lower values for [FT4] and [TSH] than what would be expected had these patients received purely L-T4 substitution therapy. This is entirely plausible because the local intracellular endoplasmic reticulum localized DIO2 (type II 5′-deiodinase) in the glial cells (e.g., tanycytes and astrocytes) of the hypothalamus and the thyrotrophs of the pituitary converts T4 into T3 locally. In the hypothalamus, FT3, directly and locally generated from deiodination of FT4 in the glia [4], is then transported to surrounding TRH neurons to suppress TRH expression. In the adenohypophysis, T3 from both circulating FT3 and that from deiodinated T4 arising directly within the TSH-secreting thyrotrophs multiply to induce TSH gene suppression.

Thus, despite the fact that T3 administration suppresses TSH, this local 5′-deiodination and the associated high intracellular concentration of T3 are chiefly responsible for the inhibitory effect on the secretion of TSH because the local intracellular concentration of T3 predominates over that contributed by plasma [FT3] [5].

Clinical data where patients were treated only with L-T4 demonstrated a clear exponential relationship. Moreover, the model that involves the observable variables of [FT4] and [TSH] has been proven to be correct. In contrast, any modeling attempt incorporating normal plasma [T3] or [FT3] has no practical utility, because both [T3] and [FT3] are products of the necessary local tissue deiodination of [FT4] and under that condition have no inhibitory impact on the secretion of [TSH] [6].

A parameterized logarithmic form of the relationship between [FT4] and [TSH] was described by Suzuki et al. [7], but the rationale behind this finding was incomplete and fundamentally incorrect because the result was supposed to describe the relationship between independent TFTs in a cross-sectional analysis of a population. A correct interpretation was described by Goede et al. [8] and introduced this relationship as applied only to intra-individual TFTs together with a clinical validation. The theoretical results include two independent HP model parameters. Then we find a parameterized expression describing the relationship between the concentrations of [FT4] and [TSH].

$$[TSH] = S \exp(-\varphi[FT4]) \tag{8.1}$$

This model has been explored further [9] where measurements' methodologies and error relationships that impact on the validity of this exponential relationship are discussed. Such considerations are fundamental for the appreciation and understanding of the necessary protocol and subsequent correct interpretation of the measured TFTs in order to provide optimum test results for the derivation of the HP characteristic [10].

In the function of (Equation 8.1), a characteristic has been described in a two-dimensional Cartesian space where parameters S and φ are real positive numbers which are extracted from measured individual TFTs.

One and only one exponential function is completely defined by the coordinates of two different points P_1 and P_2 in the [FT4]–[TSH] plane according to

$$\varphi = \left(\frac{1}{([FT4]_1 - [FT4]_2)} \right) \ln \left(\frac{[TSH]_2}{[TSH]_1} \right) \tag{8.2}$$

$$S = [TSH]_1 \exp(\varphi[FT4]_1) = [TSH]_2 \exp(\varphi[FT4]_2) \tag{8.3}$$

$$P_1 = ([FT4]_1, [TSH]_1) \tag{8.4}$$

and

$$P_2 = ([FT4]_2, [TSH]_2) \tag{8.5}$$

From the theory of differential calculus, it can be proven that every exponential function also has a unique identifiable point of maximum curvature. This unique point, when known, provides the conditions for a complete reconstruction of the exponential function it was derived from. As such, the HP function can be determined by that single specific point if this is the point of maximum curvature. The accuracy of the calculated parameters S and φ is directly related to the fitting quality of the exponential function as discussed in detail in [8] and [9]. This also relates to the definition and qualification of possible outliers [10].

8.3 Derivation of the Point of Maximum Curvature from the HP Function

The set point for euthyroid homeostasis in the HPT axis feedback loop is postulated to occur in the point of the HP function where the sensitivity for any change around this point on the curve is maximal. This postulate is based on the notion that the exponential function has one and only one point where the curvature is maximal. Maximum curvature and thus maximal sensitivity for change on the characteristic can be found according to the standard curvature theory. The radius R of the curvature K on a flat function in a defined point P is described as

$$K = \frac{1}{R} \tag{8.6}$$

We define $y = f(x) = f_1$ and then the curvature K of y (f_2) is defined as

$$K = f_2 = K = \frac{d^2y}{dx^2} \left(1 + \left\{ \frac{dy}{dx} \right\}^2 \right)^{-3/2} \tag{8.7}$$

Using the HP function:

$$[TSH] = S \exp(-\varphi[FT4]) \tag{8.8}$$

Writing [TSH] = y and [FT4] = x, we have

$$y = S \exp(-\varphi x) \tag{8.9}$$

Expanding (Equation 8.7), we find

$$K = \frac{\varphi^2 S \exp(-\varphi x)}{(1 + \varphi^2 S^2 \exp(-2\varphi x))^{1.5}} \tag{8.10}$$

For the value of x where K has a maximum, the following condition has to be met

$$\frac{dK}{dx} = 0 \tag{8.11}$$

and then

$$f_3 = \frac{dK}{dx}$$
$$= \frac{[2\varphi^2 S^2 \exp(-2\varphi x) - 1]\varphi^3 S \exp(-\varphi x)\{1 + \varphi^2 S^2 \exp(-2\varphi x)\}^{0.5}}{\{1 + \varphi^2 S^2 \exp(-2\varphi x)\}^3} \tag{8.12}$$

or

$$2\varphi^2 S^2 \exp(-2\varphi x) - 1 = 0 \tag{8.13}$$

Solving this equation results in

$$x = \frac{\ln(\varphi S \sqrt{2})}{\varphi}, \quad y = \frac{1}{\varphi \sqrt{2}}, \tag{8.14}$$

from which the first derivative at the point of maximum curvature is found to be

$$\frac{dy}{dx} = -\frac{1}{\sqrt{2}} \tag{8.15}$$

The euthyroid set point equations at the HP curve may thus be summarized as:

$$[FT4] = \frac{\ln(\varphi S \sqrt{2})}{\varphi} \tag{8.16}$$

$$[TSH] = \frac{1}{\varphi \sqrt{2}} \tag{8.17}$$

Conversely, when the coordinates of the set point are known, we can derive

$$\varphi = \frac{1}{[TSH] \sqrt{2}} \tag{8.18}$$

and

$$S = [TSH] \exp(\varphi[FT4]) \tag{8.19}$$

8.4 Theory Behind of the Physiology of the Homeostatic Control Process

In Figure 8.1, the functions of the HP f_1, the function for the curvature f_2, and the derivative of the curvature f_3 (dashed line) are depicted in one graph.

For illustrative purposes, [FT4]$_{sp}$ (set point) and [TSH]$_{sp}$ (set point) which, respectively, represent the values of [FT4] and [TSH] where the euthyroid set point occurs at 15 pmol/L and 1.0 mU/L, respectively. These set point values correspond to

$$\varphi = \frac{1}{[TSH]\sqrt{2}} = 0.707 \tag{8.20}$$

and

$$S = [TSH]\exp(\varphi[FT4]) = 40400 \tag{8.21}$$

The maximum of the curvature function f_2 determines the [FT4] value of the set point. From the derivative of f_2 to [FT4], we find f_3 as described by Equation (8.12). The dashed line characteristic of f_3 in Figure 8.1 displays two extrema, one at [FT4]$_1$ and the other at [FT4]$_2$ on either side of the intersection of the f_3 curve with the [FT4] axis coinciding with the set point value [FT4]$_{sp}$. When [FT4] increases from [FT4]$_{sp}$ toward [FT4]$_2$, the sign of

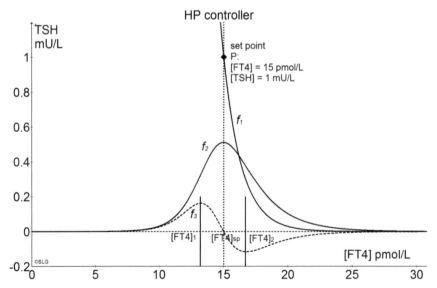

Figure 8.1   HP function f_1, curvature function f_2, and $f_3 = f_2{}'$.

f_3 is negative and prepares the HP system to increase the inhibition on TSH secretion. This inhibition reduces [TSH] which then results in a decreased thyroid hormone production rate so that [FT4] will decrease before $[FT4]_2$ has been reached and returns toward $[FT4]_{sp}$. When $[FT4]_{sp}$ drops beyond the lower level $[FT4]_1$, the HP unit detects the positive gradient of f_3 as shown by the positive sign of f_3 which then reduces the degree of TSH inhibition so as to allow [FT4] to increase toward $[FT4]_{sp}$. In this way, the HP operates as a gradient-modulated controller maintaining the value of $[FT4]_{sp}$ close to its mark.

It is useful to visualize the set point homeostatic control as a marble balanced in a bowl where its curvature is a maximum. The control of [TSH] at the set point is analogous to a marble delicately balanced on the exterior surface of the bottom of an inverted bowl resting on its circular brim on the floor. At this position of unstable equilibrium, the marble can easily roll off the bowl in either direction. This indicates that the sensitivity for change in [TSH] is highest at the point of maximum curvature. Similarly, the control of [FT4] at the set point is analogous to a marble balanced at the bottom of a bowl resting upright on its base. At this position of stable equilibrium, the marble has a strong tendency to remain where it is even if it is momentarily shifted away from this location. So, the maximum curvature of the HP characteristic is both a site where there is maximum sensitivity for [TSH] change coupled with maximum stability of maintenance of [FT4].

The value of $[FT4]_{sp}$ is derived from an internal reference that will be discussed in Chapter 9 about the HPT feedback loop.

8.5 Point of Maximum Curvature of the Thyroid Characteristic

In Chapter 7, we introduced the thyroid characteristic

$$[FT4] = A\{1 - \exp(-\alpha[TSH])\} \tag{8.22}$$

When we investigate the curvature properties of (Equation 8.22), we can derive the curvature function accordingly:

$$K = \frac{d^2y}{dx^2}\left(1 + \left\{\frac{dy}{dx}\right\}^2\right)^{-3/2} = y"\{1 + (y')^2\}^{-1.5} \tag{8.23}$$

Expanding (Equation 8.22) with the substitution of $[TSH] = x$ and $[FT4] = y$, we find:

$$\frac{dy}{dx} = A\alpha \exp(-\alpha x) \tag{8.24}$$

$$\frac{d^2y}{dx^2} = A\alpha^2 \exp(-\alpha x) \tag{8.25}$$

$$\left(\frac{dy}{dx}\right)^2 = A^2\alpha^2 \exp(-2\alpha x) \tag{8.26}$$

Then we find the curvature:

$$K = \frac{A\alpha^2 \exp(-\alpha x)}{(1 + A^2\alpha^2 \exp(-2\alpha x))^{3/2}} \tag{8.27}$$

Investigating an extremum for K results in

$$\frac{dK}{dx} = 0 \tag{8.28}$$

and then we find:

$$\{1 + A^2\alpha^2 \exp(-2\alpha x)\}^{1.5}\{-A\alpha^3 \exp(-\alpha x)\} +$$

$$- A\alpha^2 \exp(-\alpha x)(1.5)\{1 + A^2\alpha^2 \exp(-2\alpha x)\}^{0.5} \tag{8.29}$$

$$\{-2A\alpha^3 \exp(-2\alpha x)\} = 0$$

After expansion, we have

$$\{1 + A^2\alpha^2 \exp(-2\alpha x)\}^{0.5}\{2A^2\alpha^5 \exp(-3\alpha x) - A\alpha^3 \exp(-\alpha x)\} = 0 \tag{8.30}$$

or

$$A\alpha^3 \exp(-\alpha x)\{2A\alpha^2 \exp(-2\alpha x) - 1\} = 0 \tag{8.31}$$

Henceforth,

$$2A\alpha^2 \exp(-2\alpha x) = 1, \tag{8.32}$$

from which we find

$$x = \frac{\ln(\alpha A\sqrt{2})}{\alpha} \tag{8.33}$$

After the substitution of x in (8.22) with $x = [TSH]$, we get

$$y = \frac{\alpha A\sqrt{2} - 1}{\alpha\sqrt{2}} \tag{8.34}$$

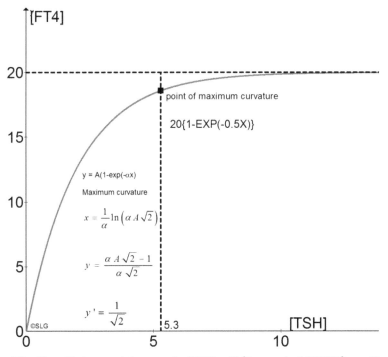

Figure 8.2   Thyroid characteristic example: [*FT4*] = 20{1 − exp(−0.5[*TSH*]} resulting in a point of maximum curvature at [TSH] = 5.3 mU/L and [FT4] = 18.6 pmol/L.

Thus, the point of maximum curvature (MK) of the thyroid is

$$[TSH]_{MK} = \frac{\ln(\alpha A\sqrt{2})}{\alpha} \tag{8.35}$$

$$[FT4]_{MK} = \frac{\alpha A\sqrt{2} - 1}{\alpha\sqrt{2}} \tag{8.36}$$

The first derivative in the point of maximum curvature is

$$\frac{d[FT4]}{d[TSH]} = \frac{1}{\sqrt{2}} = 0.707 \tag{8.37}$$

8.6 Discussion

The point of maximum curvature is a fundamental mathematical property of curves in general. A special exception is found in omni-symmetrical shaped curves [11] of which a circle is the outstanding example. With the single point of maximum curvature, the negative exponential function is particularly suited to be used as a unique point of reference for biofeedback systems.

In the following section, we will analyze this point of control reference for the closed-loop HPT system. Just as for the HP characteristic, the thyroid characteristic also has a point of maximum curvature. This point is found on a saturating exponential function and can be appreciated from the visual inspection of Figure 8.2 that for higher values of the stimulating [TSH] signal above the point of maximum curvature, the relative increase in [FT4] is strongly diminished. This property is partly compensated by the exponential growth of [TSH] at relatively lower values of [FT4], but this mechanism cannot overcome the maximum secretory capacity of the thyroid. For that reason, we can introduce the point of maximum curvature of the thyroid characteristic, as a primary marker for a clinical hypothyroid condition.

References

[1] D'Aurizio, F., Metus, P., Polizzi Anselmo, A., Villalta, D., Ferrari, A., et al. (2015). Establishment of the upper reference limit for thyroid peroxidase autoantibodies according to the guidelines proposed by the national academy of clinical biochemistry: comparison of five different automated methods. *Auto. Immun. Highlights* 6, 31–37.

[2] Koulouri, O., Moran, C., Halsall, D., Chatterjee, K., and Gurnell, M. (2013). Pitfalls in the measurement and interpretation of thyroid function tests. *Best Pract. Res. Clin. Endocrinol. Metab.* 27, 745–762. doi: 10.1016/j.beem.2013.10.003

[3] Aström, K. J., and Murray, R. M., (2009). *FeedBack Systems*, New Jersy: NJ, Princeton University Press.

[4] Gereben, B., Zavacki, A. M., Ribich, S., et al. (2008). Cellular and molecular basis of deiodinase-regulated thyroid hormone signaling. *Endocr. Rev.* 29, 898–938.

[5] Larsen, P. R. (1982). Thyroid-pituitary interaction: feedback regulation of thyrotropin secretion by thyroid hormones. *N. Engl. J. Med.* 306, 23–32.

[6] Saravanan, P., Siddique, H., Simmons, D. J., Greenwood, R., and Dayan, C. M. (2007). Twenty-four hour hormone profiles of TSH, free T3 and free T4 in hypothyroid patients on combined T3/T4 therapy. *Exp. Clin. Endocrinol. Diabetes* 115, 261–267. doi: 10.1055/s-2007-973071

[7] Suzuki, S., Nishio, S., Takeda, T., and Komatsu, M. (2012). Gender-specific regulation of response to thyroid hormone in aging. *Thyroid Res*. 5:1.

[8] Goede, S. L., Leow, M. K., Smit, J. W., and Dietrich, J. W. (2014). A novel minimal mathematical model of the hypothalamus–pituitary–thyroid axis validated for individualized clinical applications. *Math. Biosci*. 249, 1–7. doi: 10.1016/j.mbs.2014.01.001

[9] Goede, S. L and Leow, M. K. (2013). General error analysis in the relationship between free thyroxine and thyrotropin and its clinical relevance. *Comput Math. Methods Med*. 2013:831275. doi: org/10.1155/2013/831275

[10] Middelhoek, M. G. (1992). *The Identification of Analytical Device Models*. Ph.D. thesis, Delft University Press, Delft.

[11] Leow, M. K and Goede, S. L. (2014). The homeostatic set point of the hypothalamus-pituitary-thyroid axis – maximum curvature theory for personalized euthyroid targets. *Theor. Biol. Med. Model*. 11:35. doi: 10.1186/1742-4682-11-35

9

The Human Hypothalamus–
Pituitary–Thyroid Control System

"One cannot understand... the universality of laws of nature, the relationship of things, without an understanding of mathematics. There is no other way to do it."

–Richard P. Feynman (1918–1988)

9.1 Introduction

Structural analysis in physiology is of primary importance. The physiology, which displays a generalized common architecture, can be analyzed by decomposition of more or less independent parts. Although biological research seems not to be the field of engineering, the methods of fundamental research in science and technology can be applied to the research fields in the biological and biomedical disciplines. The same counts for the medical researcher who applies scientific research techniques to the life sciences and medicine to solve real-world problems such as diseases that plague mankind. Properly acquired data facilitate analysis that leads to a synthesis which elucidates biological first principles and pathways. Ultimately, a better understanding at a system level results in knowledge advancement that can be translated to the bedside to improve the treatment of patients.

As an introduction to the modeling of the homeostatic set point of the hypothalamus–pituitary–thyroid (HPT) axis, a few elementary aspects of system and function representations will be discussed. The system view of the HPT closed-loop feedback and open loop (no feedback) is given here in abstract formulations, mostly presented as mathematically modeled functional blocks which have the same meaning and properties as defined in Chapter 5 about systems theory. In this book, the HPT system is modeled from a phenomenological and functional point of view founded on physiological observations. In all instances, the physiological reality is preserved and mimicked by the model construct which is verifiable and can be supported by clinical data. The normal observable variables of the HPT system are the plasma values of free thyroxine [FT4] and thyrotropin [TSH], and sometimes [FT3]. The discussions and results in all topics of this book are aimed on real applicability on the underlying physiology and used in a clinical setting.

The value of free serum or plasma triiodothyronine concentration, [FT3], is an important HPT parameter that is sometimes measured. But its impact on the feedback loop is summed with the much higher intracellular T3 contributions arising from locally detected and converted [FT4]. Therefore, plasma values of [FT3] have no function to be incorporated in a correct model. We have to keep in mind that the measurable plasma values of [FT3] mainly (80%) reflect the 5′-deiodinase activities from active peripheral tissue conversion of [FT4]. Therefore, [FT4] is the predominant variable of interest in the control actions of the HPT loop despite this being a thyroid prohormone.

The influence on [TSH] inhibition of externally supplied T3 in the form of liothyronine tablets, or L-T3, will be discussed in Chapter 15.

Due to system complexity and practical limitations restricting observability, it is virtually impossible to take all internal effects and parameters into account. We develop here a theoretical framework which utilizes only [FT4], [TSH], and sometimes [FT3] to get a working picture of the HPT state. When we have a thorough notion of general system behavior, we can compare the effects observed from physiological systems and compare these with system behavior we know from comparable systems described in control theory. This strategy can be very effective as is shown in the following analysis where we can present a model fully with physiological realism validated by clinical data. The endocrine modeling here is done from a behavioral point of view from which the results can directly be compared with known effects from electrical network theory, signal theory, and control theory based on the integration of all physiological effects.

The complete HPT system can be divided in four major layers of control.

1. The primary control layer maintains an average level of [FT4] and is acting over longer periods of time (more than 24 h) with a distinct diurnal rhythm. This will be the main subject of this chapter.
2. The circadian charge–discharge mechanism with a 24-h rhythm will be discussed in Chapter 17.

3. The short-term feedback dynamics describes the limited oscillations of [TSH] and [FT4] as a result of their dynamic interaction with a limit cycle determined by the [TSH] half life time constant. This phenomenon will be discussed in detail in Chapter 17.

4. The fourth layer is concerned with the dynamic balance of [FT4]/[T4] and similarly the balance of [FT3]/[T3] and the related deiodinase mechanisms D1, D2, and D3 resulting in [FT4] => [FT3] and [FT4] => reverse [FT3] on tissue level [1]. Although the control mechanisms on tissue level are just as important as the previous ones, the detailed discussion falls outside the scope of this book. Partially this will be analyzed in Chapter 14, T4, T3 Half life and appearance dynamics.

The manifestation of this last layer is reflected in the [FT4]/[FT3] balance in the hypothalamus where normally a constant [FT3] is maintained. This results in a constant controlled current of [TRH] to the pituitary, which represents the value of HP model parameter S.

In this chapter about the control properties of the HPT system, we will focus on the homeostatic conditions and states of the system rather than the short-term dynamical aspects.

The analysis of dynamical physiological processes alone will not provide sufficient insight and will not result in usable applications. Therefore, we will first present a non-dynamic system behavior and analyze only the homeostatic results. After this step, we introduce time-related effects, commonly observed in the pharmacokinetic and dynamic behavior of T4 and T3. In a later stage, we will find that the normal closed-loop system is constantly oscillating with relatively small levels of variations in [FT4] and [TSH] similar to a non-linear feedback loop with hysteresis and dead time. Such a non-linear system behavior is well known from the on–off temperature controller of an ironing device.

We start by considering all system inputs and outputs as primarily non-dynamical in our discussion and analyze the system based on equilibrium or steady-state situations. A dynamical situation is analogous to an observer trying to read off values from an instrument capturing measurements of state variables that are continuously fluctuating and evolving in space and time. Only when the process has reached a state of equilibrium the "stabilized readouts" can be acquired. Still, the investigation of dynamic effects is important for a general understanding of the system.

Another important observation is related to the class of control systems of the HPT. The HPT system is a so-called single-branched control system.

This means that the growth of the [FT4] level is controlled by means of a stimulus of [TSH] to the thyroid to produce [FT4]. The reduction of the [FT4] level is accomplished only by the metabolism of T4, which results in the decline of [FT4] with an average half-life of approximately 7 days.

In the next section, we will discuss the presentation of the HPT system which we have divided into functional blocks. The function blocks themselves are modeled into two levels of signal behavior. The first one is the so-called large signal level model where the total range of input and output variables is taken into account. The second level is the so-called small signal level, valid for relatively small signal variations around a chosen point of operation on the transfer characteristic, where the steepness of the tangent at that point on the relevant curve represents the small-signal transfer factor. In other words, the small-signal transfer factor is determined by the differential quotient of the function in the point of operation.

Both signal models will be used in our analysis. In the small-signal analysis, a pronounced non-linear transfer function can be described as a linear transfer around a point of equilibrium. When time-dependent dynamic or signal behavior is involved, with time constants, delays, and frequency-dependent transfer, the model can be used in a generally transformed environment for analysis. In such a case, the Laplace transform is used for signal analysis in the complex frequency domain to avoid unnecessary differential equations. This method of modeling and model application in more complex configurations is presented as related to electrical network theory. With such a presentation, we can easily use standard electrical network simulators for further analysis.

In this chapter an elementary modeling and analysis of the HPT feedback loop will be presented without the necessity to use differential equations or Laplace transforms. In all cases, the explanation and discussion will result in simple and transparent equations fulfilling a correct presentation of the underlying physiological behavior.

9.2 Analysis of the HP Unit and Thyroid in the Closed-Loop Situation

In Chapter 5, we introduced the function blocks with a defined differential linear transfer function finally resulting in the introduction of a unity feedback system representation implying that C/R =>1 for large values of G_L.

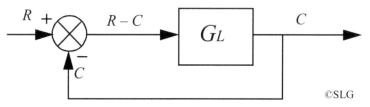

Figure 9.1 Unity gain feedback loop.

The transfer function is here presented as G_L.

This analysis is based on the differential linear behavior of the loop gain G_L [2] and has no relation with the non-linearity of the discussed separate function blocks. This implies that the fluctuations in [FT4] and [TSH] are relatively small. According to the definitions and presentations from systems theory [2], we calculate the transfer from R (reference) to C (Control) using the description of Figure 9.1.

$$C = G_L(R - C) = G_L R - G_L C, \tag{9.1}$$

resulting in

$$C + G_L C = G_L R, \tag{9.2}$$

or

$$\frac{C}{R} = \frac{G_L}{1 + G_L} \tag{9.3}$$

From Equation (9.3), we conclude that for large values of G_L, it follows that $C \Rightarrow R$, in line with the definition of a unity transfer. The overall gain factor of the feedback loop is defined as G_L.

When G_L is smaller than unity (i.e., $G_L < 1$), the amount of signal transferred from the output of the G_L block for negative feedback as input to the comparator will be too small to have any useful effect for a control loop. In such a case, the system has lost its closed-loop property and has transformed into a so-called open-loop situation. In the following sections, the subject of loop gain will be further investigated.

The HPT feedback loop can now be depicted with the use of previously introduced and analyzed function blocks. The total feedback loop of the HPT system is presented in Figure 9.2.

From Figure 9.2, we can derive the transfer from C to R which can be written as

$$\frac{C}{R} = \frac{G_1 G_2}{1 + G_1 G_2} \tag{9.4}$$

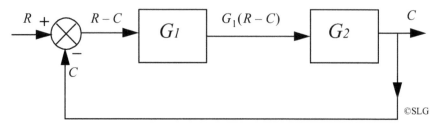

Figure 9.2   Expansion of Figure 9.1 with an additional stage G_2.

From Equation (9.4), we define the loop gain as

$$G_L = G_1 G_2 \tag{9.5}$$

This definition will be applied to the HPT feedback loop depicted in Figure 9.3.

In Figure 9.3, the internal reference value for the "final measurable" circulating [FT4] is represented by $[FT4]_R$. The output of the thyroid, with transfer factor G_T, is [FT4] and this quantity will be "sensed" by the comparator function of the hypothalamus–pituitary (HP) unit. After detection via intricate cellular signal transduction networks, the difference between the reference value $[FT4]_R$ and [FT4] (i.e., $[FT4]_R$ − [FT4]) will be amplified by G_{HP} resulting in the control signal [TSH] for the thyroid.

$$[TSH] = G_{HP}([FT4]_R - [FT4]) \tag{9.6}$$

Using Figure 9.3, we calculate the final expression for [FT4]/[FT4]$_R$ as follows. Since the output signal of the thyroid is [FT4], this leads us to the following equation:

$$G_T G_{HP}([FT4]_R - [FT4]) = [FT4] \tag{9.7}$$

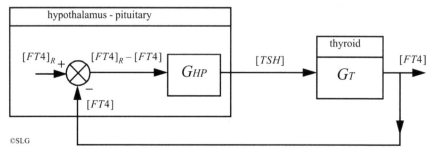

Figure 9.3   Total feedback loop of the small-signal representation of the hypothalamus–pituitary–thyroid (HPT) system.

Using the results from previous chapters,

$$G_{HP} = -\varphi[TSH], \tag{9.8}$$

and

$$G_T = \alpha A \exp(-\alpha[TSH]) \tag{9.9}$$

The modulus of the product of G_{HP} and G_T represents the loop gain factor G_L.

$$G_L = |G_{hp}G_T| \tag{9.10}$$

Expanding Equation (9.6) results in:

$$[FT4] = G_L[FT4]_R - G_L[FT4], \tag{9.11}$$

then

$$[FT4](1 + G_L) = G_L[FT4]_R, \tag{9.12}$$

thus

$$\frac{[FT4]}{[FT4]_R} = \frac{G_L}{1 + G_L} \tag{9.13}$$

The value of [TSH] is directly observable from the block diagram of Figure 9.3.

$$[TSH] = |G_{HP}|([FT4]_R - [FT4]) \tag{9.14}$$

For relatively large values of G_L, the expression of Equation (9.13) approaches 1. This implies that the [FT4] at the set point is maintained very near [FT4]$_R$, which is exactly what is expected of this control loop. Because G_L has a practical value around 4 or 5, we can see from Equation (9.13) that

$$[FT4]_R = \frac{G_L + 1}{G_L}[FT4] \tag{9.15}$$

Equation (9.15) implies that the internal reference value [FT4]$_R$ is somewhat larger than the measured set point value of [FT4] depending on the value of G_L.

For a proper and stable operation of the feedback loop, the value of G_L should always be greater than unity. In the case that G_L is smaller than unity, the output [FT4] will decrease versus the input signal [FT4]$_R$. This means that the interaction between the HP and the thyroid, T, will be lost, thereby resulting in an open-loop situation. This criterion determines the validity of the equilibrium points in the reference range as will be later discussed in detail. Here, we have to keep in mind that the preferred level of [FT4] has to be maintained as close to the set point value of [FT4] as possible. As opposed to the common notion that the thyroid status of a patient is reflected by the value of [TSH], in the context of the operational characteristics of the HPT axis, we have shown here that the controlled output value of [FT4] functions as the control criterion that determines the net integrity of the feedback loop, whereas [TSH] functions as the control signal to accomplish this target.

In this discourse of the HPT loop, it is interesting to see what will happen when the loop is interfered by an externally injected signal. This kind of attempt to analyze the closed feedback loop has been reported in various publications which showed unexpected small deviations of [FT4] [3–5] from which no concrete deductions were obtained. In some medical trials, this situation occurs when a normal person receives an external dose of T4 in the form of L-T4. This can be an anti-thyroid drug or a thyroid hormone like L-T4 as substitution for [FT4] produced by the healthy thyroid. The temporary perturbation fades away quickly and the whole system is regulated back to the normal equilibrium. Figure 9.4 shows the situation where external L-T4 is added to result in an enhanced level of [FT4] + Δ[FT4]. This enhanced [FT4] level is then in turn detected by the HP.

Let us analyze this situation in detail. For the output of the thyroid, [FT4] is augmented with an external fraction denoted by Δ[FT4]. After the integration of feedback signals at the level of the HP comparator, we find the subtracted result:

$$C_{\text{out}} = [FT4]_R - ([FT4] + \Delta[FT4]) = [FT4]_R - [FT4](1 + \Delta) \quad (9.16)$$

The input of the thyroid is then:

$$[TSH] = G_{HP}[FT4]_R - G_{HP}[FT4](1 + \Delta), \quad (9.17)$$

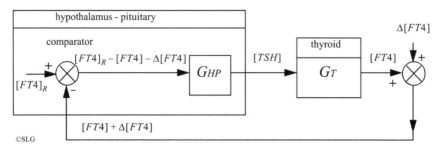

Figure 9.4   Feedback loop with an external addition of $\Delta[FT4]$.

from which we derive the output of the thyroid as follows:

$$[FT4] = [TSH]G_T = G_{HP}G_T[FT4]_R - G_{HP}G_T[FT4](1 + \Delta), \quad (9.18)$$

after which we find

$$[FT4] + G_{HP}G_T[FT4](1 + \Delta) = G_{HP}G_T[FT4]_R, \quad (9.19)$$

and then

$$[FT4] = \frac{G_{HP}G_T}{1 + G_{HP}G_T(1 + \Delta)}[FT4]_R \quad (9.20)$$

The final result, however, will be unchanged within the limits of linear operation of the feedback loop because:

$$\frac{[FT4]}{[FT4]_R} = \frac{G_L}{1 + G_L(1 + \Delta)} \quad (9.21)$$

The expression of Equation (9.21) implies that for large values of G_L, the influence of $\Delta[FT4]$ is negligible. A similar result will be obtained when the [FT4] in the bloodstream could be decreased by means of biochemical inhibition or binding, which will then lead to an increased level of [TSH].

From Equation (9.21), it is easy to conclude that the external addition of $\Delta[FT4]$ has no tangible effect on the final [FT4] result. The additional $\Delta[FT4]$ is often used to evaluate bioequivalence of different brands of L-T4. Such a transient test can reveal the individual transient responses of [TSH] which can be used for an individualized evaluation of bioequivalence of an L-T4 product. The externally supplied L-T4 will finally trigger the feedback regulation of the thyroid in an intact HPT loop to restore the [FT4] back in equilibrium. Because the temporary value of [TSH] will decrease this feedback regulation, this will also result in reduced T3 production from the thyroid. Externally administered T3 and T4 will be analyzed in Chapter 14.

9.3 Small-Signal Model of the Thyroid

In Chapter 7, the functional analysis of the thyroid was presented.

$$[FT4] = A\{1 - \exp(-\alpha[TSH])\} \quad (9.22)$$

Here, A represents the maximum value of [FT4] for large values of [TSH] and α represents the steepness to reach the saturation level. From the characteristics of the thyroid given as a saturating exponential expression, we can derive the small-signal transfer factor G_T which is defined as

$$G_T = \frac{d[FT4]}{d[TSH]} = \alpha A \exp(-\alpha[TSH]), \qquad (9.23)$$

where $G_T > 0$ for all values of [TSH].

9.4 Small-Signal Modeling of the Hypothalamus–Pituitary (HP) Unit

Without going into the details of the internal control systems with local short and ultrashort feedback loops and thyrotropin releasing hormone (TRH), we will lump the total effect of this unit in our modeling account. Literature about this subject suggested a linear logarithmic relationship between [FT4] and [TSH] [6, 7] which has been developed and validated by Goede et al. [8] of which the results were presented in Chapter 6.

The large-signal model complies to a wide range of [TSH] and [FT4] values and is presented as

$$[TSH] = \frac{S}{\exp(\varphi[FT4])} \qquad (9.24)$$

Here S is the so-called translation parameter and φ the exponential coefficient. In the same way as is shown for the thyroid, we derive the small-signal model by calculating the first derivative of Equation (9.24):

$$G_{HP} = \frac{d[TSH]}{d[FT4]} = \frac{-\varphi S}{\exp(\varphi[FT4])} = -\varphi[TSH] \qquad (9.25)$$

The set point values of $[FT4]_{\text{sp}}$ and $[TSH]_{\text{sp}}$ were published by Leow et al. [9]

$$[FT4]_{\text{sp}} = \frac{\ln(\varphi S \sqrt{2})}{\varphi}, \qquad (9.26)$$

and

$$[TSH]_{\text{sp}} = \frac{1}{\varphi \sqrt{2}} \qquad (9.27)$$

under the condition that:

$$\frac{d[TSH]_{\text{sp}}}{d[FT4]_{\text{sp}}} = \frac{-1}{\sqrt{2}} \qquad (9.28)$$

Here the reference value of the set point $[FT4]_R$ is a constant value belonging to the specific large-signal HP characteristic and is part of the internal system reference not to be confused with the set point. This distinction will be explained in the following section.

9.5 Loop Gain Analysis

The thyroid and HP unit form a closed negative feedback loop control system, maintaining a narrow area of a set point defined value of [FT4]. This control condition holds as long as the so-called loop gain G_L, which is the absolute value of the product of the thyroid transfer factor and the HP transfer factor, is greater than unity. Furthermore, from the loop gain expression, we can investigate the conditions for a loop gain optimum.

From the values of the set point, we can derive the transfer gain for the thyroid and the HP characteristic. For the thyroid transfer and HP transfer, using their belonging transfer functions, we find the expression for the HPT loop gain:

$$G_L = |G_{HP}G_T| = \varphi[TSH]\alpha A \exp(-\alpha[TSH]), \qquad (9.29)$$

where G_L is the only dependent of the variable [TSH]. Therefore, the first derivative of G_L to [TSH] results in:

$$\frac{dG_L}{d[TSH]} = \frac{\varphi\alpha A - \varphi\alpha^2 A[TSH]}{\exp(\alpha[TSH])} = 0 \qquad (9.30)$$

The optimum for G_L is then found when

$$\varphi\alpha A - \varphi\alpha^2 A[TSH] = 0 \qquad (9.31)$$

Then,

$$[TSH] = \frac{1}{\alpha} \qquad (9.32)$$

This means that we find the value for α from the set point condition:

$$\alpha = \frac{1}{[TSH]_{SP}} \qquad (9.33)$$

Substitution of α in the expression for the thyroid Equation (9.22) results in

$$[FT4]_{SP} = A\{1 - \exp(-\alpha[TSH]_{SP})\} = [FT4]_{SP}$$
$$= A\left\{1 - \exp\left(-\frac{[TSH]_{SP}}{[TSH]_{SP}}\right)\right\} \qquad (9.34)$$

With $e = 2.7182$ we find,

$$[FT4]_{\text{SP}} = A\{1 - \exp(-1)\} = 0.632A, \tag{9.35}$$

from which follows

$$A = \frac{[FT4]_{\text{SP}}}{0.632} \tag{9.36}$$

Substitution of A and α in the expression for G_L results in

$$(G_L)_{\text{max}} = \frac{\varphi A}{2.7182} = 0.41 \frac{[FT4]_{\text{sp}}}{[TSH]_{\text{sp}}} \tag{9.37}$$

The product of the model parameters φ of the HP characteristic and A of the thyroid curve determines the maximum value of the loop gain G_L. Furthermore, the properties of the HP characteristic and thus the properties of the set point determine the parameter values of the thyroid. Then the values of the set point determine the complete set of HPT properties.

Using this general formulated theory, we will illustrate the effects with the use of an average and very typical behavior of the HPT axis. For the sake of discussion, we will use a convenient commonly encountered set point [FT4] = 15 pmol/L and [TSH] = 1.5 mU/L. This set point marks the intersection of the HP and thyroid characteristics. In order to visualize this intersection properly, we have to decide in what frame of reference we want to depict the discussed effects.

1. We invert the thyroid function in order to align to the horizontal [FT4] axes or
2. We invert the HP characteristic to align to the horizontal [TSH] axes.
3. We find the inversion of [FT4] = $A\{1 - \exp(-\alpha[\text{TSH}])\}$ where we express [TSH] as a function of [FT4] resulting in

$$[TSH] = -\frac{1}{\alpha} \ln\left(\frac{A - [FT4]}{A}\right), \tag{9.38}$$

or
4. The inversion of [TSH] = $S \exp(-\varphi[\text{FT4}])$ where we express [FT4] as a function of [TSH], resulting in

$$[FT4] = \frac{1}{\varphi} \ln\left(\frac{S}{[TSH]}\right) \tag{9.39}$$

Using Equation (9.38), we find the following graphical display in Figure 9.5.

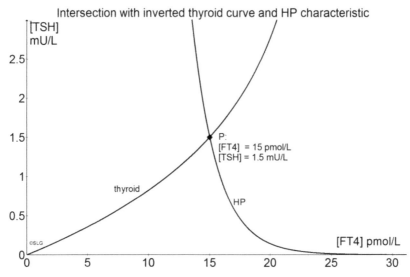

Figure 9.5 Hypothalamus–pituitary (HP) inverted thyroid intersection in [FT4] = 15 pmol/L and [TSH] = 1.5 mU/L.

The intersection of the thyroid and an inverted HP characteristic is depicted in Figure 9.6.

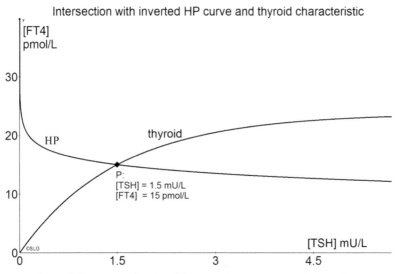

Figure 9.6 Intersection thyroid curve with inverted HP characteristic.

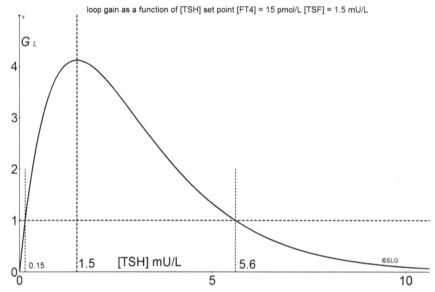

Figure 9.7   Loop gain as a function of [TSH] for set point [FT4] = 15 pmol/L and [TSH] = 1.5 mU/L. The stable range for [TSH] is indicated as $0.15 < [TSH] < 5.6$ mU/L.

The loop gain G_L, described by Equation (9.29) as a sole function of $[TSH]$ can now be depicted when the set point dependent parameters S, φ, α and A have been determined. This results in the curve depicted in Figure 9.7.

The loop gain curve with set point [FT4] = 15 pmol/L and [TSH] = 1.5 mU/L defines an operational range of [TSH] along the threshold value of G_L = 1. This demarcation line determines the individual [TSH] reference range where a stable feedback operation is defined for a person with this set point. In Chapter 10, this finding and the consequences for the traditional reference ranges will be discussed in detail.

9.6 Hypothalamus (TRH)–Pituitary (TSH) Transfer

In Section 9.2, we analyzed the properties of the general [FT4]–[TSH] feedback loop. However, there also exists a local feedback mechanism in the HP unit. The deiodinase enzyme, D2, active in the hypothalamus where [FT4] is locally converted to intracellular T3 activates local neurons responsible for the production of TRH, a stimulation hormone for the production of TSH in the pituitary. In classical thyroidology, we recognize this as the long feedback control loop eponymously known as the Fekete–Lechan loop. This process is described by Nillni et al. [3]. A publication of Prummel et al. [4] suggested even two local feedback loops. One of these loops was already known as short local feedback loop, but a second one, active locally on the pituitary itself, was identified as ultra-short feedback loop (Brokken–Wiersinga–Prummel loop). The existence of a local ultra-short feedback loop has a fundamental system control rationale. It is absolutely not desirable in a biological system that a gland can just get active without any control. This situation can be compared with the unity feedback tracking where an internally negative feedback mechanism ensures a perfect calibrated response on the stimulus. It is biologically plausible that every secretory function is equipped with such a local control mechanism. In the following, we will investigate the structure and the response of the local short feedback loop, notably the [TRH]–[TSH] response.

The short local feedback loop of the HP is depicted in Figure 9.8:

From Figure 9.8, we derive the following equations:

$$C = HPR - HPC, \tag{9.40}$$

from which we find

$$C + HPC = HPR \tag{9.41}$$

$$C(1 + HP) = HPR \tag{9.42}$$

Thus,

$$\frac{C}{R} = \frac{HP}{(1 + HP)} \tag{9.43}$$

Figure 9.9 depicts the situation of a stimulation of [TSH] by means of an externally injected amount of TRH indicated by $[\Delta T]$.

From Figure 9.9, we derive the following equations:

$$C = PH(R - C) + P\Delta T, \tag{9.44}$$

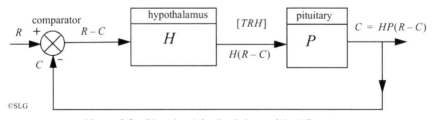

Figure 9.8   Short local feedback loop of the HP system.

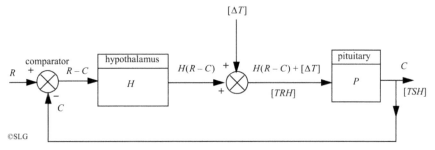

Figure 9.9   Hypothalamus–pituitary stimulation of [TSH] by means of an externally injected amount [ΔT] of injected thyrotropin releasing hormone (TRH).

resulting in

$$C = PHR - PHC + P\Delta T \qquad (9.45)$$

$$C + PHC = PHR + P\Delta T \qquad (9.46)$$

$$C(1 + PH) = PHR + P\Delta T \qquad (9.47)$$

$$C = \frac{PH}{(1 + PH)}R + \frac{P\Delta T}{(1 + PH)} \qquad (9.48)$$

From which, it is obvious that

$$\frac{P\Delta T}{(1 + PH)}, \qquad (9.49)$$

converges to zero for large values of the loop gain PH. This implies clearly that an intervention with externally injected TRH as [ΔT] has no lasting influence on the final resulting [TSH] values. Until now, we have not included the dynamic properties of the hypothalamus H and the pituitary P. These qualities can be derived from measured data responses as published by Yavuz et al. [5].

From reference [5] where two different groups of hypothyroid patients were treated as per group A treatment with L-T4 replacement and group B treatment with equivalent doses L-T3 until both attain euthyroid [TSH] values $0.4 < [TSH] < 4$ mU/L.

The study medications TRH were administered at 06:00, 12:00, and 18:00. On discharge, the patients crossed over to the alternate treatment arm following the same scheme. Study subjects underwent three doses in steps of 5, 15, and 200 µg of TRH stimulation test. TRH tests were performed after a 10-h fasting at 8:00 h in the morning and 24 h apart. The early morning

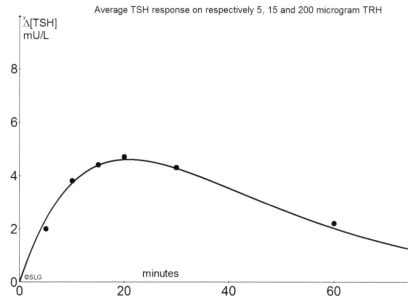

Figure 9.10 Average [TSH] response on respectively injected doses TRH of 5, 15, and 200 µg for all groups and all participants.

dose of the study medications was administered before the tests. Each dose of 200 µg/mL was directly injected in the intravenous line using a 0.5-mL (pediatric insulin) syringe, followed by a 10 mL 0.9% saline solution. There were also samples collected from a catheter at times of 15, 0, 5, 10, 15, 20, 30, and 60 min.

The responses of the [TSH] on the [TRH] infusions are plotted in the following graph depicted in Figure 9.10.

The results plotted in Figure 9.10 show a distinct pharmacodynamic impulse response that is limited in the maximum value C_{max} by the control mechanism despite the different input values of 5, 15, and 200 μg TRH.

9.7 The Detailed HPT Loop

When we combine the results from Sections 9.2 and 9.6, we come to the following schematic of the complete HPT loop depicted in Figure 9.11.

The schematic of Figure 9.11 shows a more complete HPT loop with an inner feedback path to control the intermediate [TRH] to the pituitary.

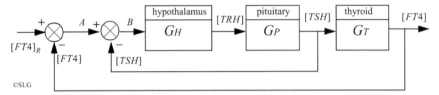

Figure 9.11   Second-order complexity of the HPT loop with inner and outer feedback paths.

9.8 Large-Signal Intercept Set Point

The feedback loop can also be presented as the physiological loop in which we will not indicate the internal reference value $[FT4]_R$ as was used in the previous feedback block diagrams. In Figure 9.12, the block diagram of the HPT physiology is depicted and the internal reference value is derived from the point of maximum curvature of the HP curve.

From Figure 9.12, we can find the equations describing the equilibrium condition for the negative HPT feedback operation. For the HP characteristic, we found without applying the extended expression.

$$[TSH] = S \exp(-\varphi[FT4]), \qquad (9.50)$$

and the transfer characteristic of the thyroid

$$[FT4] = A\{1 - \exp(-\alpha[TSH])\} \qquad (9.51)$$

These equations are interacting under the condition that Equation (9.50) is substituted into Equation (9.51), resulting in a sole implicit equation for [FT4]:

$$[FT4] = A\{1 - \exp(-\alpha S \exp(-\varphi[FT4]))\} \qquad (9.52)$$

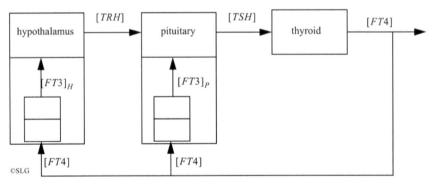

Figure 9.12   Physiological HPT feedback loop presentation.

When we substitute

$$[FT4] = x, \tag{9.53}$$

we can solve the following equation

$$x = A\{1 - \exp(-\alpha S \exp(-\varphi x))\} \tag{9.54}$$

Equation (9.54) is an implicit function of x of which we will find the solution by the following operation

$$F_1 : y = x, \tag{9.55}$$

and

$$F_2 : y = A\{1 - \exp(-\alpha S \exp(-\varphi x))\} \tag{9.56}$$

The solution for the set point is found as the intercept point of F_1 and F_2 according to

$$F_1 = F_2 \tag{9.57}$$

The proposed set point example is based on [FT4] = 15 pmol/L and [TSH] =1.5 mU/L. When we use the earlier derived set point equations, we have

$$\varphi = \frac{1}{[TSH]_{sp}\sqrt{2}} \tag{9.58}$$

$$S = [TSH]_{sp} \exp(\varphi[FT4]_{sp}) \tag{9.59}$$

$$A = \frac{[FT4]_{sp}}{0.632} \tag{9.60}$$

$$\alpha = \frac{1}{[TSH]_{sp}} \tag{9.61}$$

Then we find, respectively, $\varphi = 0.47$, $S = 1766$, $A = 23.7$, and $\alpha = 0.67$ which results in the set point equation:

$$x = 23.7\{1 - \exp(-1183 \exp(-0.47x))\} \tag{9.62}$$

When we use the template of Equations (9.55) and (9.56), we can present the solution result in Figure 9.13:

A similar expression for [TSH] can be found when we substitute Equation (9.51) into Equation (9.50). This results in:

$$[TSH] = S \exp(-\varphi A\{1 - \exp(-\alpha[TSH])\}), \tag{9.63}$$

and with the set point parameters $\varphi = 0.47$, $S = 1766$, $A = 23.7$, and $\alpha = 0.67$, and substituting [TSH] = x, we find

$$x = 1766 \exp(-11.14 + 11.14 \exp(-0.67x)), \tag{9.64}$$

resulting in the graph of Figure 9.14.

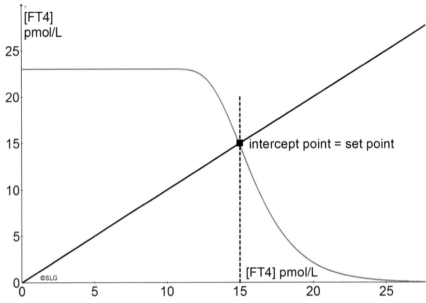

Figure 9.13   Intercept point of the large-signal representation of the physiological HPT loop.

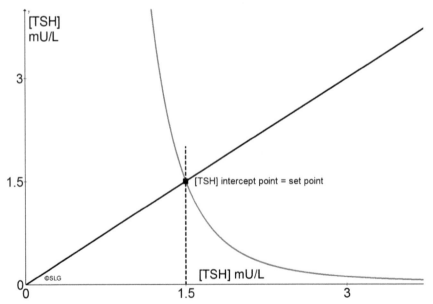

Figure 9.14   [TSH] Intercept point of the large-signal representation of the physiolog-ical HPT loop based on the set point parameters belonging to [FT4] =15 pmol/L and [TSH] = 1.5 mU/L.

9.9 Discussion

The negative feedback of a healthy HPT system ensures a constant value of [FT4] based on the internal physiologic reference $[FT4]_R$ in the HP. The HP is the main controller of the HPT system and the thyroid represents the controlled thyroid hormone production facility. One of the essential properties of a closed-loop control system is the fuzzy relationship between cause and effect. Therefore, a closed-loop system is very difficult to analyze with normal signal hormone levels. From control theory, we learn that an impulse or step response of a closed-loop system can provide all necessary information for analysis, but these methods are generally difficult to apply in the human subject because of physiological limitations and mortality risks.

In a healthy situation, the HPT system is in equilibrium and the measured value of [FT4] and the belonging value of [TSH] will be assumed to represent the euthyroid set point of the individual. With this information and the earlier discussed set point theory of Chapter 8, we can reconstruct the HP and thyroid gland parameters. This unique point defines the complete HPT system.

We found the basic functionality of the HP behavior in an open-loop situation from patients lacking a normal operating thyroid. With this validated relationship between [FT4] and [TSH], we logically inferred that the point of maximum curvature [9] was identified as the internal point of reference normally used in a closed-loop system. The introduction of the loop gain and the optimum value of this gain function established the validity of the point of maximum curvature of the HP characteristic.

With the basics of feedback systems, we have a tool to understand all kinds of control phenomena in biologic systems. It is remarkable that most bio-control systems have a robust control function and a solid stability. Despite the overt non-linearities in system parts like delay and distribution effects, the loop gain of a closed-loop system is not too high to induce such chaotic instability.

For a small part, the secretory effects and physiologic properties of the thyroid and of glands in general have been investigated. One of the prominent observations was reported by Prummel et al. [4] where an ultra-short feedback loop was discovered in the pituitary. This observation has a fundamental system rationale because a stimulated gland should have additional layers of protection from secretory over-production out of control. A local feedback tracking mechanism ensures that all glands keep on the stimulation track, i.e., the gland output is proportional to the input stimulation. This tracking mechanism has a hierarchical layering and can be found in every sub-section of a biofeedback system. Another secretory mechanism can be observed when the local tracking is not available. In such a case, the gland has a dual mode. In general, the gland will secrete just the right amount of product. As a backup, a storage reservoir can exist to be released by means of hormonal triggering or a variety of biological stimuli. An extensive overview of pulsatile secretory gland behavior is done by Veldhuis et al. [10].

References

[1] Larsen, P. R., and Zavacki, A. M. (2012). Role of the iodothyronine deiodinases in the physiology and pathophysiology of thyroid hormone action. *Eur. Thyroid J.* 1, 232–242. doi: 10.1159/000343922

[2] Aström, K. J., and Murray, R. M. (2009). *Feed Back Systems*. Princeton, NJ: Princeton University Press.

[3] Nillni, E. (2010). Regulation of the hypothalamic thyrotropin releasing hormone(TRH) neuron by neuronal and peripheral inputs. *Front. Neuroendocrinol.* 31, 134–156. doi:10.1016/j.yfrne.2010.01.001.

[4] Prummel, M. F., Brokken, L. J. S., and Wiersinga, W. M. (2004). Ultra short-loop feedback control of thyrotropin secretion. *Thyroid* 14, 825–829. doi: 10.1089/thy.2004.14.825

[5] Yavuz, S., Linderman, J. D., Smith, S., Zhao, X., Pucino, F., and Celi, F. S. (2013). The dynamic pituitary response to escalating-dose TRH stimulation test in hypothyroid patients treated with liothyronine or levothyroxine replacement therapy. *J. Clin. Endocrinol. Metab.* 98, E862–E866. doi: 10.1210/jc.2012-4196

[6] van Deventer, H. E., Mendu, D. R., Remaley, A. T., and Soldin, S. J. (2011). Inverse Log-Linear Relationship between Thyroid-Stimulating Hormone and Free Thyroxine Measured by Direct Analog Immunoassay and Tandem Mass Spectrometry. *Clin. Chem.* 57, 122–127.

[7] Suzuki, S., Nishio, S., Takeda, T., and Komatsu, M. (2012). Gender-specific regulation of response to thyroid hormone in aging. *Thyroid Res.* 5:1 doi: 10.1186/1756-6614-5-1

[8] Goede, S. L., Leow, M. K., Smit, J. W. A., Dietrich, J. W. (2014). A novel minimal mathematical model of the hypothalamus–pituitary–thyroid axis validated for individualized clinical applications. *Math. Biosci.* 249, 1–7. doi: 10.1016/j.mbs.2014.01.001

[9] Leow, M. K., and Goede, S. L. (2014). The homeostatic set point of the hypothalamus-pituitary-thyroid axis – maximum curvature theory for personalized euthyroid targets. *Theor. Biol. Med. Model.* 11:3. doi:10.1186/1742-4682-11-3

[10] Veldhuis, J. D., Keenan, D. M., and Pincus, S. M. (2008). Motivations and methods for analyzing pulsatile hormone secretion. *Endocr Rev.* 29, 823–864. doi: 10.1210/er.2008-0005

10

Reference Ranges
for TSH and FT4

"We can only see a short distance ahead, but we can see plenty there that needs to be done."

–Alan Turing (1912–1954)

10.1 Introduction

Since the advent of thyroid diagnostics, attempts have been made to decipher the key relationships governing thyroid physiology to facilitate diagnosis, treatment, and monitoring of those with thyroid diseases.

In July 1953, Sydney C. Werner wrote a research article in the Bulletin of the New York Academy of Medicine about the pituitary thyroid relationship in normal and disordered thyroid states [1]. This was one of the first attempts to describe the influence of externally administered thyroid hormone on the thyroid gland behavior. The search for the relationship between FT4 and TSH got a boost with the introduction of automated high-throughput FT4–TSH assays and the application of statistical methods. The goal was to find serum concentrations of TSH and FT4 of the sampled healthy individuals deemed to have normal thyroid status representative of the human population. This was the initial effort at defining the normal ranges of FT4 and TSH. We have also to keep in mind that the early measurement technology to establish TSH and FT4 concentrations was very primitive in the 1970s compared to the highly sensitive and specific assay technology we have at our disposal today. In this chapter, we will discuss the current situation of the reference ranges for free T4 concentrations, [FT4], and TSH concentrations, [TSH].

10.2 Reference Ranges for [FT4] and [TSH]

In terms of the healthy population, the typically encountered normal reference ranges of [FT4] and [TSH] are about $10 < $ [FT4] $ < 20$ pmol/L and $0.4 < $ [TSH] $ < 4$ mU/L, respectively. These reference ranges are largely based on statistical normalization procedures applied to TFT data acquired on a large sample size and are generally valid and applicable to guide clinical decisions.

In 2009, Ross et al. [2] showed a method of analysis regarding these bivariate reference ranges of [FT4] and [TSH] in the form of scatter plots with [FT4] on the horizontal axis and [TSH] on the vertical axis. For diagnostic applications, the relevant reference ranges of [FT4] and [TSH] as determined by the clinical chemistry lab are then used by the clinicians. It is no exaggeration that the methodology of screening of healthy subjects to establish the normal population ranges of TFT and the caveats to be considered when applying these "normal ranges" to a patient population often eludes medical doctors. Over the years, this has partly contributed to a heated debate about what constitutes a true normal range, particularly for [TSH] in which the upper limit of normal has been gradually shifted downward [3]. Even the concept of a "normal range" is being challenged as both doctors and patients continue to be bewildered and baffled over the discordance between the attainment of euthyroid status as defined biochemically by TFTs and clinically by thyroid-specific symptoms and quality of life.

The following examples illustrate this commonly encountered dilemma.

1. A hypothyroid person "A" is being treated with levothyroxine hormone replacement. Follow-ups with serial TFTs periodically over a few months later showed the [TSH] = 1.50 mU/L. According to his specialist, his [TSH] fits perfectly in the euthyroid reference ranges for [TSH], so he was sent home with the belonging dose indication of levothyroxine to sustain this [TSH] value. After some time, he comes back to the clinic with complaints about fatigue and many other symptoms that point in the direction of hypothyroidism. A further determination of $[FT4]$ results in 12.3 pmol/L and the conclusion of the clinician is that this value completely conforms to the reference range guidelines and was thus in order. Nevertheless, the complaints persisted.

2. A hypothyroid person "B" being treated with levothyroxine hormone replacement. Follow-ups with serial TFTs periodically over a few months later showed the $[TSH] = 1.50$ mU/L. According to his specialist, his [TSH] fits perfectly in the euthyroid reference ranges for [TSH],

so he was sent home with the belonging dose indication of levothyroxine to sustain this [TSH] level. After some time, he comes back to the clinic with complaints about agitation and many other symptoms that point in the direction of hyperthyroidism. A further determination of [FT4] results in 18.8 pmol/L and the conclusion of the clinician is that this value completely conforms to the reference range guidelines and was thus in order.

This seems puzzling on the surface because both persons have been titrated to a median value [TSH] of 1.5 mU/L well within the region of the "normalized" TFT reference range defined as biochemical euthyroidism. A plausible explanation is that they have both been titrated to the wrong set point (i.e., the incorrect combination of [TSH] and [FT4]). The theory of the euthyroid homeostatic set point has been discussed in Chapter 8 where the individuality of the position of the specific [FT4]–[TSH] combination is indicated in the [FT4]–[TSH] plane. The examples of person "*A*" and "*B*" are depicted in Figure 10.1.

From Figure 10.1, one can appreciate the different equilibrium positions of person "*A*" located at [FT4] = 13 pmol/L and [TSH] = 0.8 mU/L while that of person "*B*" at [FT4] = 17 pmol/L with [TSH] = 2.5 mU/L.

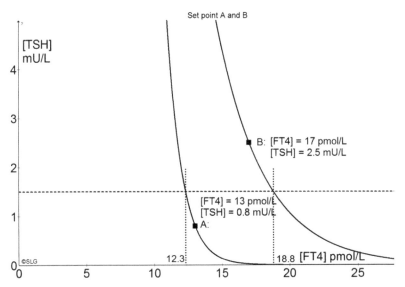

Figure 10.1 Set points of person "*A*" and person "*B*" who were titrated to [TSH] = 1.5 mU/L.

Accordingly, person *A* has a relative deficiency of T4 while person *B* is overdosed. This occurs when the levothyroxine doses are titrated to achieve a $[TSH] = 1.5$ mU/L as this is a clinically accepted normal value for [TSH].

Without the application of the individual homeostatic set point position calculated from the available [FT4]–[TSH] combinations of previous TFTs, the clinician has no knowledge about what TFTs constitute optimal individualized euthyroid set points to render any wise judgment on the best TFT to be targeted. When the set point of a person is established from the previous TFTs during levothyroxine treatment to the euthyroid area, we have all the information that normally characterizes the HPT system. From the results of Chapter 9, we can evaluate the properties of the loop gain G_L to illustrate the diversity in euthyroid ranges that belong to a certain set point.

10.3 Set Point Determined Euthyroid Ranges for [TSH]

The value of the homeostatic equilibrium, defined as the set point in a normal person, determines the individualized [TSH] range for a proper operating feedback loop (i.e., the HP system and the thyroid can influence each other). This mode of operation is maintained as long as the earlier defined loop gain G_L is larger than one. At the moment the loop gain declines to $G_L < 1$, the HP and thyroid will no longer exert any influence on each other, thereby resulting in an open-loop situation. Such a situation is exemplified by either an overactive thyroid which produces excessive thyroxine resulting in hyperthyroidism or an underactive thyroid that secretes insufficient thyroid hormones resulting in hypothyroidism.

In the following, the expression for the loop gain G_L will be expanded and investigated for $G_L > 1$. This will result in an individualized range for [TSH] that ensures a normal feedback operation. Such a personalized approach defines the healthy [TSH] limits for which a diagnosis of disease will be made only if [TSH] should fall beyond the boundaries on this individualized range rather than outside the laboratory-defined population range.

The loop gain G_L of a normal operating HPT system is defined as

$$G_L = \left| \frac{\alpha A \varphi [TSH]}{\exp(\alpha [TSH])} \right| \tag{10.1}$$

Notably, G_L is a sole function of [TSH].

We find an extremum for G_L at:

$$\frac{dG_L}{d[TSH]} = \frac{\alpha A\varphi(1 - \alpha[TSH])}{\exp(\alpha[TSH])} = 0 \tag{10.2}$$

The extremum value of G_L is found for

$$1 - \alpha[TSH] = 0, \quad \text{or} \quad [TSH] = \frac{1}{\alpha}, \text{and after substitution}: \tag{10.3}$$

$$G_{\exp HP} = \frac{\alpha A\varphi[TSH]}{\exp(\alpha[TSH])} = \frac{A\varphi}{e}, \tag{10.4}$$

with $\exp(1) = e = 2.7182$

For the exponential thyroid model, we then find:

$$[FT4]_{SP} = A\left\{1 - \exp\left(-\frac{[TSH]_{SP}}{[TSH]_{SP}}\right)\right\} = A\{1 - 1/e\} = 0.632A \tag{10.5}$$

$$A = \frac{[FT4]_{SP}}{0.632} \tag{10.6}$$

From the previous discussion, we derived the maximum loop gain value of G_L which coincided with the set point value for [TSH]. When we plot the loop gain as a function of [TSH], we find, together with the demarcation criterion of $G_L > 1$ for a stable HPT closed-loop operation, the following result as depicted in Figure 10.2.

Table 10.1 provides a more precise indication of the indicated [TSH] ranges.

In many cases we can use the simple, but correct criterion to establish the HPT feedback condition with a single TFT as follows.

$$0.41\frac{[FT4]}{[TSH]} > 1 \tag{10.7}$$

When the expression of (10.7) < 1 we encounter an open loop situation and further action for diagnostics and treatment has to be implemented.

10.4 Different Reference Ranges in Different Laboratories

In Section 10.2, we discussed the reference ranges for [FT4] and [TSH], conveniently chosen here as $10 < [FT4] < 20$ pmol/L and $0.4 < [TSH] < 4$ mU/L which are close to the typical population ranges used by laboratories.

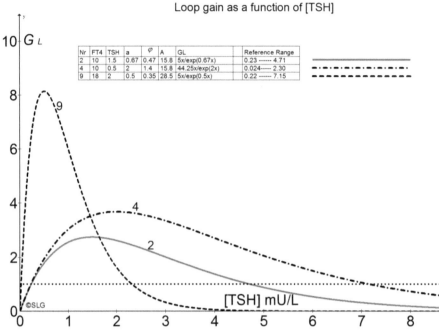

Figure 10.2   Loop gain curves of G_L as a function of $[TSH]$. The belonging curve inter-section with the line $G_L = 1$ shows the upper and lower levels of the individual euthyroid $[TSH]$ range.

Table 10.1   Set-point-related [TSH] ranges

Nr	[FT4]	[TSH]	a	φ	A	G_L	[TSH] Range mU/L
2	10	1.5	0.67	0.47	15.8	5x/exp(0.67x)	0.23–4.71
4	10	0.5	2	1.4	15.8	44.25x/exp(2x)	0.024–2.30
9	18	2	0.5	0.35	28.5	5x/exp(0.5x)	0.22–7.15

In a clinical environment, this form of "one size fits all" reference ranges established for the population is commonly accepted as the standard. In the period after 2005, the majority of laboratories all over the world use a one-step direct analog immunoassay (DAI) for the determination of free thyroid hormone. In a review about the current status of these assays, Welsh and Soldin [4] gave a detailed report about the properties and anomalies to be expected when these methods for the determination of [FT4] are applied for routine TFTs. The main rationale behind the application of these direct immunoassays is cost consideration. The time necessary to give a test result is considerably less than the time normally necessary for the preparation of the free (unbound) fractions.

For a long time, the equilibrium dialysis (ED) method for the preparation of the free fraction was considered to be the golden standard in clinical laboratories. However, the time necessary for this analyte preparation was considerable. New assays for [FT4] and [FT3] were introduced, skipping this time-consuming ED step which resulted in higher values of the measured [FT4] and [FT3] compared to the ones done with ED. This urged clinical laboratories to redefine their reference ranges for [FT4] and [FT3]. Because different manufacturers produce a slightly different DAI, this results in variations in published reference ranges. All current DAI methods for the determination of [FT4] and [FT3] also deviate from the long accepted "log-linear relationship" between [TSH] and [FT4], or, as we alluded to, the HP characteristic.

In various publications of van Deventer, Jonklaas, and Soldin [4–7], this deviation is addressed, and as an alternative for an accurate determination of the free thyroid hormone fraction, either ED or ultra-filtration is recommended in combination with an analysis step with tandem mass spectrometry equipment. This alternative may hopefully become the preferred methodology in the future when tandem mass spectrometry is widely adopted as the standard for the determination of free thyroid hormone. In the following, we will discuss the consequences of the different reference ranges commonly applied in different hospitals.

Normally, the immunoassays for the determination of [TSH] do not generally differ so widely in their measurement results as to give rise to direct concern. With the examples of three different commonly used [FT4] clinical reference ranges based on DAI methods, the consequences for the interpretation of [TSH] values are demonstrated below.

In Figure 10.3, we depict the situation of a set point value at $[TSH]_{sp} = 1.5$ mU/L.

From Figure 10.3, we appreciate the clear differences in the relationships between [FT4] and [TSH]. The only common denominator shared by the three HP curves is the value of $[TSH]_{sp} = 1.5$ mU/L. When a clinic works with such differences in the [FT4] reference range arising from different DAI kits chosen by the clinical lab in question, the resulting TFTs will lack robustness as a reliable diagnostic parameter. From this example, we appreciate that every defined [FT4] reference range has a belonging set point value for [FT4] together with the mentioned $[TSH]_{sp} = 1.5$ mU/L.

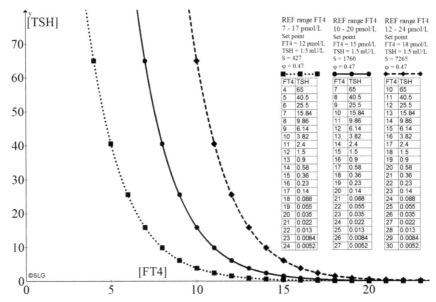

Figure 10.3 Three different HP characteristics with the same set point value of $[TSH]_{sp} = 1.5$ mU/L. The solid drawn characteristic marked with the round points indicates an HP characteristic based on a $10 < [FT4] < 20$ pmol/L reference range implying a median value of $[FT4] = 15$ pmol/L. The dotted HP curve with the square points is based on $7 < [FT4] < 17$ pmol/L implying a median value of $[FT4] = 12$ pmol/L and the dashed HP curve marked with diamond points represents the results based on $10 < [FT4] < 24$ pmol/L as the reference range for $[FT4]$ implying a median value of $[FT4] = 18$ pmol/L.

10.5 Discussion

Until today most clinicians involved in the diagnostics process of thyroid disorders only use the laboratory results of a TSH measurement. The [FT4] is only determined when the value of [TSH] falls outside the reference range of the clinical laboratory. This is the way clinicians have been educated about this subject and has to be taken into account when a medical engineer discusses these matters with a clinician.

Although the clinician is responsible for the correct treatment of a thyroid disorder based on reliable TFTs, there are inherently linked factors related to laboratory errors and assays variability, and relevant knowledge in biochemistry, physiology, and mathematics of the HPT axis, which are often left out of the diagnostic process. This increases the probability of erroneous interpretations of laboratory results by clinicians and clinical chemists which can potentially harm the patient. This analysis shows the responsibility gaps and knowledge gaps that can lead to suboptimal or even flawed treatment.

In the analysis of the set-point-related optimal loop gain, the euthyroid range for [TSH] was established by the condition that the individual loop gain has to be greater than unity for a stable feedback operation. This condition is directly derivable from the values of [FT4] and [TSH] based on the individual set point. Because the set point directs to the area of optimal HPT loop operation, it is also a marker for boundaries that indicates where the system is not operational anymore. In the latter scenario, the thyroid and HP lose their mutual interactive connection at the moment the loop gain is less than unity.

The previous discussions and analysis about the HPT set point were based on the application of a standardized reference range for [FT4] defined as $10 < $ [FT4] < 20 pmol/L. With the introduction of DAIs to establish the free fraction of [T4], we encounter the situation of multiple products from multiple suppliers for this analysis method. This results in the current differences of [FT4] reference range definitions in different clinical laboratories and even within the same laboratories that opt to switch to different assay vendors. As has been demonstrated, these differences make it difficult if not impossible to conduct a reliable analysis of TFT results.

The other major element in this discussion is represented by the determination of [FT4]. In the literature, "gold standards" like ED and currently the ultra-filtration followed by liquid chromatography and tandem mass spectrometry (LC/MS/MS) have been described for the correct measurement of [FT4] [4–7]. However, such standardization is still far away from realization. The application of the low-cost and fast DAIs for [FT4] is very popular in many clinical laboratories all over the world, but the quality and the reproducibility of the measured results remain questionable. Because every DAI for the determination of [FT4] is somewhat different, depending on the assay properties of the supplier, the laboratory that uses these establishes their own [FT4] reference range by using their standard sample of healthy subjects and then finding the belonging median and associated standard deviations as boundaries for the local [FT4] reference range. The analysis about the quality of DAIs [8] could help to find better solutions.

10.6 Concluding Remarks

Reference ranges for [FT4] and [TSH] are crucial to the clinical treatment of thyroid patients. With the application of current theoretical knowledge about the HPT system, the use of population reference ranges can be considered obsolete. We still have to keep in mind that a correct establishment of the HP model parameters, which are fundamental for the set point and accurate diagnostics, can be realized only with a well-defined, standardized, and calibrated measurement system for [FT4], [FT3], and [TSH] to allow individualized reference ranges to be utilized at the bedside for the benefit of the patients.

References

[1] Werner, S. C. (1953). Pituitary thyroid relationship in normal and disordered thyroid states. *Bull. N. Y. Acad. Med.* 29, 523–536.

[2] Ross, H. A., den Heijer, M., Hermus, A. R., and Sweep, F. C. G. (2009). Composite reference interval for thyroid-stimulating hormone and free thyroxine, comparison with common cutoff values, and reconsideration of subclinical thyroid disease. *Clin. Chem.* 55, 2019–2025.

[3] Laurberg, P., Andersen, S., Carlé, A., Karmisholt, J., Knudsen, N., and Pedersen, I. B. (2011). The TSH upper reference limit: Where are we at? *Nat. Rev. Endocrinol.* 7, 232–239. doi: 10.1038/nrendo.2011.13

[4] Soldin, S. J., Soukhova, N., Janicic, N., Jonklaas, J., and Soldin, O. P. (2005). The measurement of free thyroxine by isotope dilution tandem mass spectrometry. *Clin. Chim. Acta* 358, 113–118.

[5] Jonklaas, J., Soldin, S. J. (2008). Tandem mass spectrometry as a novel tool for elucidating pituitary-thyroid relationships. *Thyroid* 18, 1303–1311.

[6] Serdar, M. A., İspir, E., Ozgurtas, T., Gulbahar, O., Ciraci, Z., Pasaoglu, H., et al. (2015). Comparison of four immunoassay analyzers for relationship between thyroid stimulating hormone (TSH) and free thyroxine (FT4). *Turk. J. Biochem.* 40, 88–91. doi: 10.5505/tjb.2015.65487

[7] Welsh, K. J., and Soldin, S. J. (2016). How reliable are free thyroid and total T3 hormone assays? *Eur. J. Endocrinol.* 175, R255–R263. doi: 10.1530/EJE-16-0193

[8] Soldin, O. P., and Soldin, S. J. (2011). Thyroid hormone testing by tandem mass spectrometry. *Clin. Bioichem.* 44, 89–94. doi: 10.1016/jclinbiochem.2010.07.020

11

Extrapolation to the Set Point from a Single Available Measurement

The way out is through the door. Why is it that no one will use this method?
The right path is never straight, is it left curved or upwards down?
Follow the path until the right place has been found.

–Confucius (551 BC–479 BC)

11.1 Introduction

In Chapter 6, we introduced the methods and results of a mathematical formulation of the relationship between [FT4] and [TSH]. This relationship has been identified as an HPT axis characteristic of an individual. Such a euthyroid set point is the result of genetic programming and unique constitutional makeup of the individual and the associated endocrine system. Every measured pair of [FT4] and [TSH] belongs to that specific curve which represents the collection of all possible [FT4]–[TSH] equilibrium states. It manifests itself as the [FT4]–[TSH] path on which we certainly find the point of maximum curvature on which the set point is found. Based on this assumption, we know that when we have only one coordinate [FT4]–[TSH] and we assume that this point belongs to the [FT4]–[TSH] path, we can calculate the presumptive set point value using a reasonable conjecture to a certain average [TSH] or [FT4].

The following explanation will provide further insights.

11.2 Calculation of [FT4] Assuming a Pre-determined Set Point Value of [TSH]

In some cases, it happens that a person is suffering from a hyperthyroidism which results in relatively high values of [FT4] with a belonging [TSH] level that is barely detectable.

Furthermore, it can be assumed that the [TSH] reading in such a case becomes inaccurate and loses a lot of significance. The standard procedure to treat this patient is the application of anti-thyroid drugs like carbimazole, etc., in order to reduce the thyroidal production of T4 and T3. After a new equilibrium situation of several weeks, a new TFT will be performed and the new [FT4]–[TSH] status is evaluated. In many cases, there will be some degree of overshoot to low levels of [FT4] < 10 pmol/L with a belonging significantly elevated [TSH] > 10 mU/L. From that thyroid status, the patient will receive a reduced dose of carbimazole as part of a decremental regimen to down-titrate to the more euthyroid ranges of [FT4] and [TSH].

In this new situation, we have now a reliable measurement of [FT4] and [TSH] that belongs to the HP characteristic of the person in question. In Figure 11.1, we see an example of what we call "the average HP characteristic." This HP curve has a set point of [FT4] = 15 pmol/L with a belonging [TSH] = 1.6 mU/L.

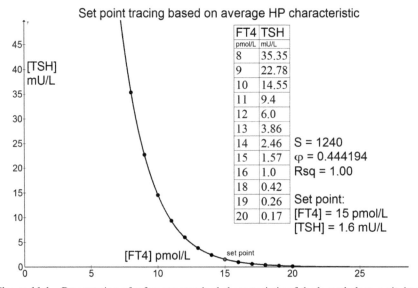

Figure 11.1   Presentation of reference standard characteristic of the hypothalamus pituitary.

Suppose we have measured [FT4] = 8 pmol/L with a belonging [TSH] = 35.35 mU/L, then we could extrapolate along the curve in such a way that we could determine the value of [FT4] if we assume that the belonging average [TSH] is about 1.6 mU/L.

This is derived with the following equations.

$$[TSH] = S \exp(-\varphi[FT4]) \tag{11.1}$$

The set point is defined according to

$$\varphi = \frac{1}{[TSH]\sqrt{2}} \tag{11.2}$$

$$S = [TSH]\exp(\varphi[FT4]) \tag{11.3}$$

These results can be used to estimate the position of the set point based on a single TFT measurement. When we assume the general [TSH] median value of 1.6 mU/L of a reference range, $0.4 < [TSH] < 4$ mU/L, we can calculate the position of the set point as follows. The coordinates of the measured point are $[FT4]_1$ and $[TSH]_1$. This implies that the value of S can be found according to:

$$S = [TSH]_1 \exp(\varphi[FT4]_1) \tag{11.4}$$

In order to complete the calculation, we have to know the value of φ.

This value can be found from

$$\varphi = \frac{1}{[TSH]_{\text{target}}\sqrt{2}} \tag{11.5}$$

where $[TSH]_{\text{target}}$ represents the [TSH] value we estimate the occurrence of the targeted set point. Using the general expression of the HP characteristic as described by Equation (11.1), we can write

$$[TSH] = \frac{S}{\exp(\varphi[FT4])} \tag{11.6}$$

We find

$$[TSH]_{\text{target}} = \frac{S}{\exp(\varphi[FT4])}, \tag{11.7}$$

resulting in

$$\exp(\varphi[FT4]) = \frac{S}{[TSH]_{\text{target}}}, \tag{11.8}$$

thus

$$\varphi[FT4] = \ln\left(\frac{S}{[TSH]_{\text{target}}}\right), \tag{11.9}$$

and the wanted value of $[FT4]$ is found by

$$[FT4] = \frac{1}{\varphi}\ln\left(\frac{S}{[TSH]_{\text{target}}}\right), \tag{11.10}$$

or

$$[FT4] = [TSH]_{\text{target}}\sqrt{2}\ln\left(\frac{S}{[TSH]_{\text{target}}}\right), \tag{11.11}$$

because φ is known, the value of S can be calculated as shown in (Equation 11.11) which results in

$$[FT4] = [TSH]_{\text{target}}\sqrt{2}\ln\left(\frac{[TSH]_1 \exp(\varphi[FT4]_1)}{[TSH]_{\text{target}}}\right), \tag{11.12}$$

and thus

$$[FT4] = [TSH]_{\text{target}}\sqrt{2}\ln\left(\frac{[TSH]_1 \exp(\varphi[FT4]_1)}{[TSH]_{\text{target}}}\right), \tag{11.13}$$

resulting in

$$[FT4] = [TSH]_{\text{target}}\sqrt{2}\left(\ln\left(\frac{[TSH]_1}{[TSH]_{\text{target}}}\right) + \varphi[FT4]_1\right), \tag{11.14}$$

with

$$\varphi = \frac{1}{[TSH]_{\text{target}}\sqrt{2}} \tag{11.15}$$

This results in

$$[FT4] = [FT4]_1 + [TSH]_{\text{target}}\sqrt{2}\ln\left(\frac{[TSH]_1}{[TSH]_{\text{target}}}\right) \tag{11.16}$$

depending on

$$\frac{[TSH]_1}{[TSH]_{\text{target}}}, \tag{11.17}$$

which can have a value smaller than one or larger than one.

When $\frac{[TSH]_1}{[TSH]_{\text{target}}}$ is less than 1, the value of $\ln\left(\frac{[TSH]_1}{[TSH]_{\text{target}}}\right)$ is negative.

From the previous theory, we can give an overview of the result we get when we use three different measured [FT4]–[TSH] values and three different targeted [TSH] values for the extrapolated set point.

In Table 11.1, we present the results of this approach.

Table 11.1   Set point extrapolation from single measurements with target levels of [TSH] = 1, 1.5, and 2 mU/L

Measured Point [FT4]	Measured Point [TSH]	[TSH] Target 1 mU/L [FT4]	[TSH] Target 1.5 mU/L [FT4]	[TSH] Target 2 mU/L [FT4]
10	25	14.5	15.96	17.4
10	16	13.9	15.02	15.88
10	8	12.94	13.55	13.92

11.3 Calculation of [TSH] Assuming a Set Point Value of [FT4] = 15 pmol/L

In this example, we can use the expression from (Equation 11.16) resulting in

$$[FT4]_{\text{median}} = [FT4]_{\text{measured}} + [TSH]_x \sqrt{2} \ln \left(\frac{[TSH]_{\text{measured}}}{[TSH]_x} \right) \quad (11.18)$$

Then we find

$$[TSH]_x \sqrt{2} \ln \left(\frac{[TSH]_{\text{measured}}}{[TSH]_x} \right)$$
$$= |[FT4]_{\text{measured}} - [FT4]_{\text{median}}| = \Delta[FT4] \quad (11.19)$$

Then

$$\frac{\Delta[FT4]}{[TSH]_x \sqrt{2}} = \ln \left(\frac{[TSH]_{\text{measured}}}{[TSH]_x} \right) \quad (11.20)$$

This is an implicit equation with the variable $[TSH]_x$.

The solution of this equation can be found graphically as is shown in the example of Figure 11.2. We can use the numbers from the HP characteristic represented in Figure 11.1. Using the hypothyroidism condition [FT4] = 9 pmol/L with the belonging value of [TSH] = 22.78 mU/L, we can write assuming [FT4] median = 15 pmol/L.

According to Equation (11.20), we find $\Delta[FT4]$ = 15 – 9 = 6 pmol/L.

The left part of the equation is then

$$FL = \frac{\Delta[FT4]}{[TSH]_x \sqrt{2}} = \frac{6}{[TSH]_x \sqrt{2}} \quad (11.21)$$

This function is depicted as the dashed curve in Figure 11.2.

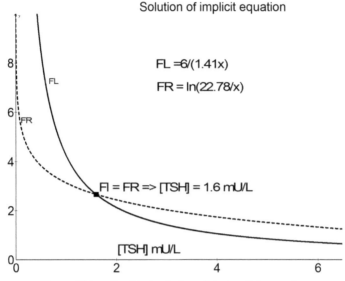

Figure 11.2   Graphical solution of the implicit equation.

Furthermore, we have the right-hand part of the equation

$$FR = \ln\left(\frac{[TSH]_{\text{measured}}}{[TSH]_x}\right) = \ln\left(\frac{22.78}{[TSH]_x}\right),\tag{11.22}$$

depicted as the blue curve in Figure 11.2.

11.4 Extrapolation to a Set Point Using a "Phantom" [FT4]–[TSH] Point

In the previous section, we used presumptive targeted values of [TSH] or [FT4] deemed likely to represent common euthyroid values consistent with optimal health. This approach suffers from the weakness that not everyone who is stably euthyroid and healthy has either a [FT4] of 15–16 pmol/L or [TSH] of 1.5–1.6 mU/L. We began with the premise that a personalized approach to elucidating the euthyroid set point requires using a set of reliable TFT data of the individual concerned, with the proviso that the fitting reliability or R^2 is dependent on the quantity and the quality of data points available. This is obviously the ideal situation. But in reality, we generally encounter patients who either possess a bunch of TFTs done by different labs (thereby introducing variability and errors) or very few TFT laboratory reports such that we may then be forced to use only a couple of the most reliable points to reconstruct the euthyroid set point.

When one encounters a large set of TFT results in which only a fraction of the full data set is useable to compute the set point, with most [TSH] values below 0.1 mU/L, then we need to apply a strategy of data selection for the best guess. In such situations, we choose TFT data points spread over the range of thyroid status with the highest values of [TSH].

It should be noted that [TSH] values larger than 100 mU/L are just as valuable as the ones in the range of $0.1 < $ [TSH] $ < 100$ mU/L because these values can be measured with adequate accuracy. As we will discuss in Chapter 12, the reliability and accuracy of [TSH] values below 0.1 mU/L are suboptimal and hence might well be left out for a reliable data set to reconstruct the HP characteristic. Importantly, under no circumstances, should one choose a bunch of TFT data points that are spread narrowly apart as this minimizes the R^2 (goodness-of-fit or coefficient of determination) in terms of the parameters defining the exponential function of best fit which increases the confidence of an accurately predicted euthyroid set point.

We now discuss a situation where we have only one solitary TFT value (obviously better if the [TSH] $ > 4$ mU/L for maximum accuracy), and we wish to compute an approximate euthyroid set point with which to target toward useful strategy when making an informed decision to select appropriate therapeutic drug doses. This may occur when a clinician sees a new patient who has only one TFT record for the very first time. A TFT done at that initial clinical encounter may then be potentially useable to compute the set point using another method that minimizes any bias from assuming that the set point [FT4] or [TSH] should be at any particular pre-determined values.

This method introduces the concept of a so-called "phantom" [FT4]–[TSH] position. In this case, we postulate a theoretically exaggerated supranormal value of [FT4] = 100 pmol/L with an associated [TSH] suppressed to a practically undetectable value. This is largely true as almost more than 99% of humans with normal HPT systems who become so thyrotoxic will have profound TSH suppression. We will discuss the impact of two phantom points based on a targeted set point value of [TSH] = 1.5 and [TSH] = 1 mU/L and show the effect of the differences. We will use Figure 11.3 as a template for our discussion.

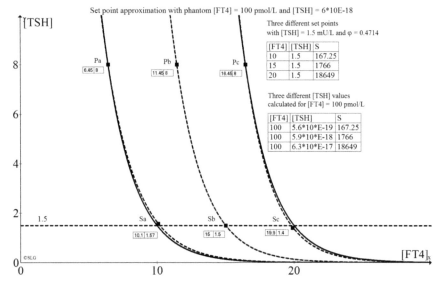

Figure 11.3   For the phantom point, we choose [FT4] = 100 pmol/L. From an HP character-istic with a set point of [FT4] = 15 pmol/L and [TSH] = 1.5 mU/L, we obtain the belonging value for [TSH] = 5.9×10^{-18} mU/L.

The phantom point has to be constructed based on the properties of one or more real-life HP characteristics. This implies that the phantom [TSH] value, belonging to the chosen phantom [FT4] of 100 pmol/L, is a number depending on the HP characteristic function.

In the following, we choose three different starting points Pa, Pb, and Pc as indicated in Figure 11.3 as possibly available for the clinician to start with. When we have chosen the target [TSH] set point level to be 1.5 mU/L, we can construct three curves.

1. From Pa [FT4] = 6.45 pmol/L [TSH] = 8 mU/L to the set point Sa [FT4] = 10.1, [TSH] = 1.57.

From Table 11.2, we appreciate that the phantom value of [TSH] is a tiny number which differs considerably per dedicated curve.

Table 11.2 also indicates that we have to choose the correct value of [TSH] belonging to the phantom point; otherwise, it will not be on the targeted HP curve. In this example, we choose the phantom point to be [FT4] = 100 pmol/L and [TSH] = 5.9×10^{-18} mU/L. Then we have the median black curve of Figure 11.3.

Table 11.2 Set point target [TSH] = 1.5 mU/L and phantom [FT4] = 100 pmol/L and $\varphi = 0.4714$

Starting Point [FT4]	Starting Point [TSH]	Set Point [FT4]	Set Point [TSH]	S	Phantom [TSH]
6.45	8	10	1.5	167.25	5.6*10E − 19
11.45	8	15	1.5	1766	5.9*10E − 18
16.45	8	20	1.5	18649	6.3*10E − 17

Next, we construct HP curves with Pa and Pc, respectively, with the phantom point.

From Table 11.3, we appreciate that the set point values based on the phantom point and starting points Pa and Pc do not differ significantly from the original ones.

In the next step, we will explore the effects of a targeted set point [TSH] of 1 mU/L and compare the results of the set points in Table 11.3. This is depicted in Figure 11.4. We use the same starting points, Pa, Pb, and Pc, and find different curves indicated as dashed orange, dashed green, and dashed purple. The set point values for [FT4] with [TSH] = 1 mU/L have been altered very minimally.

In Table 11.4, we present the related data based on a targeted [TSH] set point value of 1 mU/L.

From Figure 11.4, we appreciate that the resulting set point values with phantom point [FT4] = 100 pmol/L and [TSH] = 5×10^{-27} mU/L have a neglectable difference compared to the original set points. Furthermore, we can see that the differences compared to the [TSH] = 1.5 set point values are just about 0.6 pmol/L. The results of the discussed examples then justify the application for practical clinical purposes for a first rough indication the phantom point of [FT4] = 100 pmol/L with a [TSH] = 6×10^{-18} mU/L.

Table 11.3 Phantom point [FT4] = 100 pmol/L and [TSH] = 5.9×10^{-18} mU/L

Starting Point [FT4]	Starting Point [TSH]	Set Point [FT4]	Set Point [TSH]	Phantom [TSH]
6.45	8	10.1	1.57	5.6*10E − 19
16.45	8	19.9	1.4	6.3*10E − 17

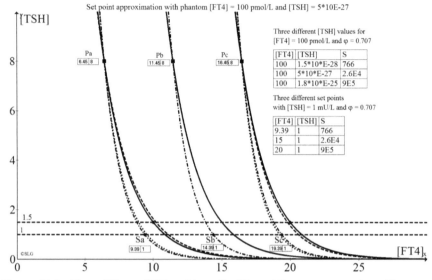

Figure 11.4   Three different TFT positions, Pa, Pb, and Pc, as an example to illustrate a possible HP characteristic based on a phantom point with [FT4] = 100 pmol/L where the belonging value of [TSH] is calculated based on the presumed set point values for [TSH] = 1.5 and [TSH] = 1 mU/L.

Table 11.4   Set point trajectories with different phantom points

Starting Point [FT4]	Starting Point [TSH]	Set Point [FT4]	Set Point [TSH]	S	Phantom [TSH]
6.45	8	9.4	1	766	1.5*10E − 28
11.45	8	14.4	1	2.6E4	5*10E − 27
16.45	8	19.4	1	9E5	1.8*10E − 25

11.5 Discussion

We have illustrated how one can predict a "reasonable" euthyroid set point from a single reliable measurement of [FT4] and [TSH]. We can estimate the possible set point value of [FT4] when we assume that the belonging set point value of [TSH] could be about 1.5 mU/L. Depending on the initial measured TFT, we next calculate the belonging possible value of [FT4]. A similar procedure can be followed to estimate a possible value of [TSH] based on the assumption that the possible set point value for [FT4] is about 15 pmol/L. However, in this case, we have to find the solution with graphical methods which can be easily accomplished using plotting software. These estimations are based on the previously discussed theoretical frame work.

A third approach that provides acceptable set point positions is based on the assumption of a phantom second point which is positioned at [FT4] = 100 pmol/L and a [TSH] of 6×10^{-18} mU/L. The idea is of the phantom point is that the [TSH] should have the correct value to present acceptable approximated HP curves.

The graphical presentations of Figures 11.3 and 11.4 give an idea about the dispersion of the extrapolated set points where the numerical results are given in Tables 11.2–11.4.

12

Measurement Methods, Error Sources, and Error Interpretation of FT4 and TSH Concentration Values

"An experiment is a question which science poses to Nature, and a measurement is the recording of Nature's answer".

– Max Planck (1858–1947)

12.1 Introduction

For scientific disciplines such as physics, chemistry, and systems biology, computations and the interpretation of numerical information is vital. Archimedes, in the early days of science, noticed an increase in water level when he entered the bathtub. This led to his momentous discovery which stemmed from an astute observation rooted in the capacity to quantify and express changes related to various factors in terms of numbers. During the ancient Egyptian era when the pyramids were built, reliable and standardized methods of measurement of length and weight evidently existed. The Egyptians could accomplish amazing historical achievements because they diligently applied the internal standards of measurement and calibration. In old documents, we learn that standards used by ancient civilizations contributed to their societal sophistication and advancement. Take, for instance, the measurement standards for length such as the biblical cubit = 45.72 cm or Stadium = 184.9 m. And for volume, Choinix = 1.09 L, whereas for weight, the Sikkel = 11.43 g or a Talent = 41.15 kg. From the Greeks, we also know that they formalized and refined a measurement system with a mathematical foundation based on geometry.

A monumental event in Western European history was the French revolution of April 1789. This marks the beginning of a new era with the introduction of the metric system in 1795. For any measurement system, a defined calibrated standard is a fundamental condition. Every culture has some kind of measurement system in order to handle the practical daily issues of change, amount, and comparison with a defined reference standard. In modern science, we have a standardized system of measurement units based on the metric system, namely: *Le Système International d'Unités* or SI for short. The SI defines seven fundamental physical units of measure from which every other unit can be derived. These are the meter for length, the kilogram for mass, the second for time, the kelvin for temperature, the ampere for electrical current, the candela for luminous intensity, and the mole for amount of substance. When we have established our measurement references and standards, we can find out to what extent a certain measured value is actually true. Some thoughts about truth and reality are indicated as quotes in the beginning of this page, but the notion of truth is central when we discuss the meaning, interpretation, and impact of measurements.

The precision of routine automated immunoassays such as thyroid function tests (TFTs) is generally assumed unquestionably reliable and robust for clinical decision making with respect to diagnosis and treatment. However, there is presently no universal consensus on what constitutes acceptable standards for the measurement of free thyroxine [FT4] and thyrotropin [TSH]. This issue will be discussed in a later paragraph.

In this chapter, we examine the nature and sources of common errors and deviations that are encountered in measurement methods of plasma FT4 and TSH based on the hypothalamus–pituitary (HP) characteristic that was discussed in Chapter 6. The HP relationship between [FT4] and [TSH] [1, 2] is particularly relevant in the modern era of personalized medicine where inaccuracies of [FT4] and [TSH] can compromise individualization of patient-specific treatment endpoints based on mathematical models for homeostatic set point determination [3]. The results of a formal theoretical analysis of these errors will enable chemists, biologists, clinicians, and medical scientists to get an understanding of the underlying mechanisms and how to achieve a proper measurement interpretation.

Current literature reveals that a negative exponential model fits clinical data best with respect to the relationship between [FT4] and [TSH] and is thus far the most tenable and enduring of all existing theoretical constructs [1]. Former publications using scatter plots of homeostatic values of hundreds of measured [TSH]–[FT4] combinations from different individuals [4–6] hinted at a "logarithmic [TSH]-linear [FT4]" relationship although based on erroneous regression analyses. Such scatter plots are two-dimensional representations of a certain distribution of [TSH], coupled with a belonging [FT4] value. It is helpful to remember that the normal reference ranges of [TSH] and [FT4] are established by each clinical laboratory via the statistical distribution of [TSH] and [FT4] of a group of healthy subjects from the general population. In these scatter plots, every coupled [TSH]–[FT4] coordinate represents the current value of the points of equilibrium of an individual thyroid status in the investigated population. Since the physiological processes and parameters of each individual are unique like a fingerprint and different people can thus have widely separated positions of [TSH]–[FT4] pairs on such plots despite having similar thyroid hormone status, it is impossible from these published scatter plots and their related regression analyses to draw any valid conclusion about the [TSH]–[FT4] relationship within any given person across the spectrum of thyroid hormone status.

However, the scatter plots have additional qualities which can provide an unexpected insight in modeling. For instance, there are persons with nearly similar euthyroid set points despite different HP physiological processes and parameters. There are also people with distinctive deviations of the normal set point albeit with relatively similar HP characteristics. The HP characteristic curve should therefore be considered as the collection of all possible points of equilibrium of an individual. Some are more in the direction of high [FT4] with a low [TSH], whereas others are more in the direction of low [FT4] and relatively high [TSH]. In all cases, we find a distinct negative exponential relationship.

The proposed parameterized negative exponential model derived from earlier published results [1] and validated with individual series of TFTs here is:

$$[TSH] = \frac{S}{\exp(\varphi[FT4])} \tag{12.1}$$

The model has two degrees of freedom, S and φ. These degrees of freedom open the possibility to position the graph of the model anywhere on a two-dimensional Euclidean plane. The model can be validated with measurements of [FT4]–[TSH] pairs of persons with overt primary hypothyroidism treated with L-T4 over a number of months till their [FT4] and [TSH] are fine-tuned to the coordinate space surrounding their original pre-disease euthyroid set point.

Biological variations are expected to occur in [FT4] and [TSH]. A very important condition for a TFT measurement is the time of day when a blood sample is taken as many hormone systems in the body exhibit natural circadian bio-rhythms dependent on time, including the HPT axis. The diurnal rhythm of [TSH] plays an important role and exhibits significant differences between morning and evening readings [7].

The amount of variability is dependent on the individual in question, but the smallest inter-individual variations in [TSH] are observed around 15.00 h in the afternoon [7].

In another study [8], it was evident that several persons being probed for [FT4] after taking their daily dose of L-T4 had different readings because of inter-individual differences in T4 appearance and metabolism. Therefore, it is important to probe a person already using L-T4 at a fixed time of the day before the intake of the daily L-T4 dose. For practical reasons, this can be done shortly upon awakening (i.e., prior to the ingestion of daily dose of L-T4) in the early morning between 07.00 and 10.00 h. The same time interval for TFT assessment also applies to people being investigated for the first time.

12.2 Error Definitions and Error Sources

Definition: An error value is the difference between measured result and the real value.

A common practice in the presentation of measurement results is the rounding of a certain number resulting in an absolute error. At least two forms of rounding procedure can be recognized.

1. Rounding to the nearest integer above the value of the measurement $M+0.5$ to $M+1$ or to the nearest lower integer beneath $M+0.5$ to M. This method results in an uncertainty of ± 0.5.
2. Rounding with the use of truncation of the decimal value leaving behind the primary integer. For example, 5.2 can be truncated to 5, but 5.9 can also be truncated to 5. This method leaves an uncertainty of ± 1.

The second definition is the relative error defined as the relative difference between measured value and true value.

$$\varepsilon = \frac{M - \upsilon}{\upsilon},$$

(12.2)

where M denotes the value of the measurement, υ represents the true value, and then ε represents the relative error.

3. Measurement errors as a consequence of an imprecise direct analog immunoassay for the determination of [FT3] and [FT4] [6] because of incomplete separation of bound and unbound fractions.

4. Measurement differences as a result of different measurement methods and different standards [6] Every laboratory has the freedom to choose their preferred test equipment.

12.3 Measurement Theory

In this discussion, we will make a distinction between measured results and the phenomena from which these results are obtained. When we observe a certain phenomenon, we could try to find out what this really could be. By means of a theory, we could find an explanation. A very important example of an observed phenomenon can be found in the history of astronomy. Although length measurement was developed to a solid method establishing the value of length, the application of measurements was from a knowledge theoretical point of view based on a mathematical framework and correctly applied by the Greek mathematician Eratosthenes to determine the value of the earth contour. Because he observed the properties of shadows of long vertical poles on various places on a large area, he theorized that the earth had to be a sphere and that the contour of the earth could be calculated by means of the length of the poles and the related length of the shadows.

We can appreciate two separate components:

1. an established method for measurement technology;
2. development of a theory based on careful observations in which the measurement technology could be applied.

These two elements of a measurement system and a theory are fundamentally necessary conditions to come to a finalized result based on a theoretical framework. A third step has to be taken to establish the validity of the theoretical results. This step belongs to the stage of validation [9]. According to this example, the real value of the earth contour could be determined with nautical measurement units and today with satellite technology.

When the validation has been established in one or more independent ways, we have the assurance of the validity of our theory and/or our measurements. These examples imply that measurements are only relevant when the person who uses these measurements knows what has to be measured and why! Measuring is meaningful as long as you know what you are measuring. Unfortunately, this turns out not to be the case in various instances of medical practice.

A related issue central to the topic of errors in measurement is the manner of statistical analysis as commonly applied in medicine and related biomedical and psychosocial disciplines. Ironically, these are the very fields of investigation concerning very complex phenomena of which formalized research and mathematical modeling have hardly started to define the simplest ideas. Data analysis with inferential statistics may seem conceptually sound based on probabilistic notions at its core and is thus widely applied. Yet, it suffers from drawbacks because such an approach often does not yield a theoretical framework that explains the underlying mechanisms of the biological processes in question. It cannot be overemphasized that we actually need to understand why we apply a certain mathematical technique because otherwise the entire exercise may well end up as "garbage-in, garbage-out" type of readouts which makes no sense. Truth lies not in all data *per se* but may be wrested out reliable data acquired from scientifically rigorous methodologies and analyzed using appropriate techniques. The correct use of statistical methods in physics is well established from the theoretical frameworks of statistical mechanics, thermodynamics, plasma physics, and quantum mechanics. In signal theory, the mathematical method of stochastic variables and signals is a solid example.

As we have seen in the examples of Chapters 4 and 6, the statistical methods currently applied in medical and clinical research, such as applying a regression line through a data cloud of unrelated measurements, are unlikely to generate any real underlying biological insights. On the other hand, the use of descriptive statistics in medical research remains a time honored method to explore the possible expected measurement sample space. A valuable example can be found in the presentation of TFTs in the [FT4]–[TSH] plane. This will show a cloud of dots with the highest density centered around the median value of [FT4]–[TSH] pairs. This density will decrease when we go to the peripheral values of this plane.

In the following sections, we will focus on methods and interpretations of the most common measurements in thyroidology – the values of [FT4] and [TSH]. Also, the consequences of deviations in measurement results will be analyzed.

12.4 [FT4] and [TSH] Measurement or Thyroid Function Tests (TFTs)

The theoretical framework derived from observations and theorizing is based on observations of the thyroid system. Essentially, we have established and validated our theory on the variables that can be observed and measured.

Although [T3], [FT3], and many other related hormone levels can be measured, we have determined that [FT4] and [TSH] are the dominant variables of our model which was validated in Chapter 8. Because the relationship between [FT4] and [TSH] has to obey the expression of (Equation 12.1), we calibrate all the measurements we will discuss with the HP model of Chapter 6. This means that at the moment a series of TFT measurements deviate from that relationship, we have possibly acquired erroneous data.

It has been demonstrated to be very important that the TFT is performed on a defined time of day under specific circumstances. In order to minimize all possible noise and disturbances, all TFTs should preferably be done between 7.00 and 10.00 h in the morning before taking the daily dose of levothyroxine and breakfast. The average time interval between TFT's is about 6 weeks. This interval assures a new equilibrium when the clinician has checked the results and possibly alters the dosage.

12.5 Sensitivity of Model Parameter φ as a Function of [FT4] and [TSH]

An in-depth study on the deviation effects and errors on [FT4] and [TSH] measurements has been published [2] from which we will discuss the main results. The relative deviations are fairly constant over the [FT4] range of $5 <$ [FT4] < 25 pmol/L. For values of [FT4] < 5 pmol/L, the relative deviations in φ increase steeply in an exponential-like manner, thereby implying a strong increasing deviation and measurement uncertainty for φ at overtly hypothyroid levels of [FT4] that will necessarily result in less reliable outcomes to construct a true HP function. For values of [FT4] < 5 pmol/L, the accuracy deteriorates rapidly.

12.6 Measurement Errors from [FT4] Rounding or Truncation Procedure

The following discusses the effects of approximation by "rounding" or even "truncation" of decimals of measurements in a [FT4]–[TSH] plot.

Figure 12.1 indicates how a variation of 9–10 pmol/L results in a variation of [TSH] varying from 21 to 13.6 and an overall difference of 7.4 mU/L, while a variation in [FT4] from 15 to 16 pmol/L results only in a [TSH] difference of 0.6 mU/L.

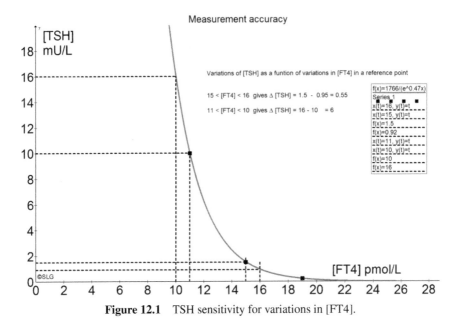

Figure 12.1   TSH sensitivity for variations in [FT4].

This shows the importance of accurate measurements for lower values of [FT4], while the readings for higher values of [FT4] are less sensitive for variations in [TSH].

Many clinical and biochemistry laboratories follow a practice of expressing [FT4] values to the first decimal place. There are also laboratories which either round off the value of [FT4] to the nearest integer or truncate away decimals and leave behind only the integer. This problem surfaces because the laboratory scientists and the physicians in this present day and age cannot communicate sufficiently due to the separation of knowledge, technical capabilities, and responsibilities to achieve efficiency in the current medical organization. Clearly, the measurement methods are neither designed nor performed by medical personnel charged with the ultimate responsibility for the evaluation of the measured data.

As an example, the laboratories state that their choice of truncating data is basically a practice to avoid misleading clinicians to over-interpret trivial changes in serial results which may actually reflect analytical and biological "noise." Equally, there are also laboratories which are similarly erroneous by using the default number of decimal places that appear on the analyzer printout without making adjustments to match the analytical imprecision of their assays. The assay issue will be discussed at the end of this chapter.

Furthermore, it shows the effects in a certain part of the HP characteristic when the rounding procedure is performed on one single [FT4] measurement resulting in a rounded integer. This means that the maximum rounding error can be defined as [FT4] $\pm$ 0.5 pmol/L. The consequences for the error magnitude depend strongly on the position on the curve. For example, a measured value of [FT4] = 11.4 pmol/L will be rounded down to 11 pmol/L, while a measured value of 11.6 pmol/L will be rounded to up to 12 pmol/L. In the first case, we will find a difference in [TSH] of 6 mU/L. In the second instance, the error will be 3.75 mU/L. This rounding error will decrease for the higher ranges of [FT4], provided that the [FT4] assay is based on equilibrium dialysis or LC tandem mass spectrometry [6].

12.7 Absolute Deviation of [FT4] as a Function of Relative [TSH] Error

Using the expression of [FT4] and [TSH] of Equation (12.1), we will investigate the deviation effects on [FT4] as a function of the relative error in [TSH]. Suppose the relative error in [TSH] is represented by $p\%$, then the upper limit value of [TSH] is

$$UL_{[TSH]} = [TSH](1 + p/100) \qquad (12.3)$$

and the lower limit will be

$$LL_{[TSH]} = [TSH](1 - p/100) \qquad (12.4)$$

This has the following effect on the related values for [FT4].

We can write [FT4] as

$$[FT4] = \frac{1}{\varphi} \ln\left(\frac{S}{[TSH]}\right) \qquad (12.5)$$

The absolute difference in the related values for [FT4] as the upper limit of [FT4] is

$$\Delta[FT4] = \frac{1}{\varphi}\left(\ln\left(\frac{(1 + p/100)}{(1 - p/100)}\right)\right) \qquad (12.6)$$

This means that the absolute error of [FT4] is a function of the relative error, p, of [TSH] and φ and is independent of S and is also constant over the entire range of [FT4].

This is a direct consequence of the exponential properties of the relationship between [FT4] and [TSH].

Table 12.1   Illustration of absolute error [FT4] with different relative errors in [TSH] and different values of φ

φ	$p = 1\%$ Δ[FT4] pmol/L	$p = 2\%$ Δ[FT4] pmol/L	$p = 5\%$ Δ[FT4] pmol/L
0.3	0.066	0.13	0.33
0.4	0.05	0.1	0.25
0.5	0.04	0.08	0.2
0.6	0.033	0.07	0.17
0.7	0.028	0.06	0.14
0.8	0.025	0.05	0.13

12.8 The Effect of Absolute Errors in [FT4] Resulting in Deviations of [TSH]

The absolute error of plus or minus λ in the values of [FT4] can be expressed as:

$$F_{([FT4]+\lambda)} = \frac{S}{\exp\{\varphi([FT4] + \lambda)\}} \tag{12.7}$$

and

$$F_{([FT4]-\lambda)} = \frac{S}{\exp\{\varphi([FT4] - \lambda)\}} \tag{12.8}$$

The resulting absolute error in the value of [TSH] can be written as:

$F_\varepsilon = F_{([FT4]-\lambda)} - F_{([FT4]+\lambda)}$, resulting in

$$F_\varepsilon = \frac{2S\sinh(\varphi\lambda)}{\exp\{\varphi[FT4]\}} \tag{12.9}$$

The effects of "λ" are shown as [FT4] $\pm$ 0.5 compared to the results when $\lambda = 0.1$, as can be appreciated from Table 12.2.

When λ is reduced from 0.5 to 0.1 the reduction in deviations of [TSH] is quite obvious. For a more accurate measurement result of [TSH] as a function of [FT4], it is as we have to present [FT4] with a decimal number instead as a whole integer. Because this method is justified as derived mathematically

Table 12.2   The effect of absolute error $F\varepsilon$

[FT4] $\pm$ 0.5	$F\varepsilon$	[FT4] $\pm$ 0.1	$F\varepsilon$
15	$\pm$0.3	15	$\pm$0.05
10	$\pm$2.9	10	$\pm$0.45
5	$\pm$25	5	$\pm$4

above, laboratories should also be cognizant of their [FT4] accuracy level using appropriate immunoassay platforms in order to comply with a well understood necessity of [FT4] accuracy.

12.9 Errors from Physiological Memory Effects, or Hysteresis

We learned from clinical experience that euthyroid set points are expected in a certain area of the reference ranges of [FT4] and [TSH]. However, in cases of strong hyper- or hypothyroid conditions, the disordered physiology needs a significant amount of time to restore the system to normal operation. When a severe hyperthyroidism occurs, the overwhelmingly elevated levels of [FT4] introduce an excess of thyroid hormone load on the HP unit and result in a disturbance of the [FT4]–[TSH] relationship, which needs some weeks to months for recovery toward the normal HP response curve. In Figure 12.2, this effect is indicated in the dashed curve showing that the normal HP characteristic is shifted to the lower end of the [FT4] axis and is reflected by a relatively long suppressed [TSH]. A similar effect is found in people with a prolonged severe hypothyroidism with high levels of [TSH]. Such cases may also take some time before the memory of this condition has been removed for the normal HP response to return, as indicated by the blue dashed curve implying that the HP curve is shifted to the higher end of the [FT4] axis.

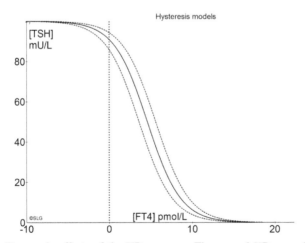

Figure 12.2 Hysteresis effects of the HP response. The normal HP curve is indicated as drawn black, between the dashed and dotted curves.

This memory effect, in which a system is dependent not only on its current state but also on its former environment, is recognized as hysteresis and is a phenomenon well characterized in many areas of physics as has been observed and analyzed in the magnetization curves of ferromagnetic materials. In physiology and medicine, hysteresis has also been encountered in pulmonary mechanics, bladder smooth muscle stretch, and even calcium-parathyroid hormone relationship [10–12]. With thyroid physiology, hysteresis causes a shift of the original characteristic over the horizontal axis, or a shift in the value of parameter S, which is equivalent to translation along the [FT4] axis. When the HPT axis encounters sustained elevated concentrations of [FT4] well above the extreme of the upper normal limit (i.e., thyrotoxicosis), it is consistently observed that the recovery response of [TSH] lags behind that of [FT4]. It takes considerable time for definitive treatment of the thyrotoxicosis before [TSH] will be detectable again even when [FT4] declines to subnormal levels. Evidently, the system "remembers" the huge amount of [FT4] and needs time to remove the effects. This is actually a beneficial adaptive feature that is evolutionarily conserved to protect the organism in case [FT4] should suddenly escalate rapidly without restraint, since the [TSH] will remain low and not add undue further stimulation to the thyroid. When the [TSH] response is detectable again, we see a shift of the original HP characteristic to the lower end of the range of the [FT4] scale.

A similar effect is observable after a long standing hypothyroid condition with [FT4] concentrations depressed far below the lower normal limit. In this case, the [TSH] normalization response lags behind the recovery of [FT4] such that [TSH] may still remain in the supra-normal levels despite [FT4] having achieved normal or even relatively high levels due to the HP curve being shifted to the higher range of the [FT4] values.

In the case of an endured period of thyrotoxicosis where the amount of T4 and T3 is more than a super physiological concentration, the hypothalamus is adjusted to an enhanced internal equilibrium of [FT3], resulting in a reduction of the TRH current to the pituitary. In fact, the model parameter S is down modulated to a reduced value. We encounter this effect as a shift of the HP characteristic to the lower end of the [FT4] scale. During the time this state is continued, we find a lowering of the [FT4] set point value. Similarly, when a prolonged state of hypothyroidism is maintained, the equilibrium of [FT3] in the hypothalamus is reduced, resulting in an enhanced TRH current to the pituitary resulting in a higher value of model parameter S, which results in a shift of the HP characteristic to the higher end of the [FT4] scale.

Obviously, this hysteresis represents a protective mechanism as the persistently elevated [TSH] provides a "buffer" of additional thyroid stimulation in case [FT4] should suddenly diminish to negligible from an unpredictable loss of exogenous T4 supply.

12.10 Errors from Biochemical Assay Technology for [TSH] and [FT4]

The previous discussions about the relationship of [FT4] and [TSH] are based on the fact that changes in [FT4] result in exponential variations in [TSH]. This phenomenon has been widely validated with multiple TFT values of one person over a wide range of [TSH]. TSH and FT4 concentrations are measured with specialized equipment and methods of which the measurement methods of [TSH] have a good level of standardization widespread over all laboratories in the world. The detection and measurement of [FT4], until the year 2000, was commonly performed with the separation method for bound and free T4 known as "equilibrium dialysis" (E.D) or the Nichols gold standard method. This measurement standard provided accurate and reproducible results and established the log [TSH]-linear [FT4] relationship.

Thyroxine (T4) is about 99.96% bound to carrier proteins in the blood circulation [2]. This protein binding renders the bound T4 fraction inactive, while the free fraction of T4, [FT4] is biologically active on the HP system and responsible for the overall regulation of thyroid hormones in the HPT axis. Binding concentrations of T4 are dependent on individual health status and have a lower level in disease. In special healthy cases like pregnancy, thyroxine binding proteins can be elevated to 50%, but the normal HPT control system ensures a sufficient level of [FT4]. The [FT4] fraction is therefore the only clinically meaningful variable to be determined.

Today, the most common method for the measurement of [FT4] in clinical laboratories is the so-called direct analog immunoassay. In this method, a separation step of bound T4 and free T4 is not performed and the free T4 fraction is more or less estimated by the measurement result. This method for the determination of the [FT4] fraction leads to a flawed measurement result because there is no clear determination of the free T4 fraction in the first place. Therefore, we find a pollution of bound T4 fractions in the result. Furthermore, the measurement results are highly variable when performed again on the same blood sample. Therefore, direct [FT4] immunoassays are fraught with errors. An alternative is liquid chromatography (LC) tandem mass spectrometry following a separation step for the bound and free fraction of T4.

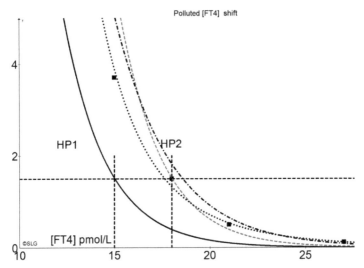

Figure 12.3   Clean [FT4] measurement after separation of bound T4 and free T4 indicated with the black HP1 curve. The dashed HP curve is the result of polluted analyte with bound fractions of T4 and results in an upward shift of [FT4] from 15 to 18 pmol/L.
The square dots indicate the non log - lin compliance.

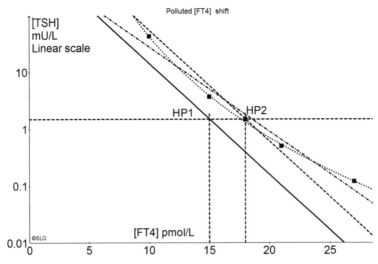

Figure 12.4   In this figure, the [TSH] axis is presented as logarithmic.
The square dots indicate the non log - lin compliance.

The steps to separate the free fraction of [T4] can be either equilibrium dialysis or ultrafiltration which correlates well with the results of the equilibrium dialysis Nichols method as published in [4]. Via this method, the correlation with the log [TSH]-linear [FT4] relationship was clearly confirmed [5]. The direct analog immunoassays for [FT4] results in a positive median shift of the [FT4] reference range from about 15 pmol/L to the a new median of about 18–19 pmol/L for the reference ranges defined by the suppliers of direct immunoassays. In Figures 12.3 and 12.4, we show the effects of currently popular direct [FT4] immunoassays and compare the bound T4 pollution effects with real established (LC MS/MS) measurement methods for [FT4].

An unpolluted set of [FT4] measurements would result in a straight line with a logarithmic [TSH] scale, such as the black dashed lines. The dotted line represents the 100% fit on a power series data set, but results in a poor exponential fit as indicated in the dashed lines.

12.11 Discussion

In the overview above, we presented a variety of error sources that can be encountered in [FT4]–[TSH] measurements. In order to reconstruct the HP characteristic of an individual in a reliable way, we have to make sure that the effects of extreme situations like long standing hypothyroid or hyperthyroid conditions are also taken into account. Furthermore, a standardized time of the day (e.g., early morning between 07.30 and 09.30 h) for blood sampling for [FT4]–[TSH] measurement is necessary. A good knowledge and control of the laboratory measurement accuracy together with a more accurate representation of [FT4] values up to at least an extra decimal digit is highly recommended.

Over the years, the measurement technology and accuracy has improved tremendously. This means that measurements in the previous decade or earlier (e.g., years 2000–2003) are probably less accurate than the ones in 2012–2013. Preferably, the measurement results of one and the same laboratory with the same calibration accuracy and methodological technique are used to determine the HP characteristic. It is important to realize that the approximation of [FT4] assay results is an operationally standardized protocol related to the presentation of laboratory results in certain laboratories, but it is necessary to appreciate the imprecision caused in terms of the error that can arise when such results are applied to the automated tools for the reconstruction of the HP characteristic. In order to prevent unnecessary [FT4] shifts to the higher

end of the [FT4] range, clinical laboratories have to standardize their measurement methods and calibrate [FT4] measurements to either equilibrium
dialysis and/or LC tandem mass spectrometry [13, 14]. This will ensure the
conformity to the log [TSH]-linear [FT4] relationship.

There are of course other error sources like the monthly hormonal changes
in pre-menopausal women across the menstrual cycle which can influence the
TBG effects on [T4] [15]. When all error sources and effects are taken into
account, we can use the laboratory assay values of [TSH] and [FT4] with the
coefficient of variations to determine the accuracy of the predicted calculated
HP curve.

Besides the mentioned enhanced TBG effects we have to mention the fundamental changes in the physiology of a pregnant woman. During gestation
the HPT system is adjusting to the situation that the embryo needs an adapted
dosage of thyroxine. The first three months result in a an enhanced level of
human Chorionic Gonadotropin or hCG which activates together with [TSH]
the thyroid to an enhanced production. Also the value of model parameter S
is adjusted by a larger TRH current to the pituitary. After three months we
encounter a shift of the HP characteristic over an extra 3 pmol/L to the higher
end of the [FT4] axis.

12.12 Conclusion

In order to reconstruct a reliable HP function, the following error sources
in TFT determination have to be taken into account including the recommended measurement regimens. We can distinguish some sources to be more
dominant in their effects than others.

1. For general TFT measurement, the circadian variations in TSH are
   dominant in the results.
2. With the determination of [FT4] of persons using L-T4, for practical
   reasons, the blood sample is taken (preferably in the morning) before
   the daily dose of L-T4 has been administered and without breakfast.
3. For a consistent reconstruction of the HP characteristic, only TFT values
   from the same lab should be used where the [FT4] values are measured
   with methods based on equilibrium dialysis, and/or tandem mass spectrometry. Immunoassays for [FT4] without a separation of bound T4 and
   [FT4] are advisably avoided if practicable.
4. Determination and reporting of [FT4] should preferentially have an
   accuracy of up to at least one decimal value (± 0.5 pmol/L).

5. The possible deviation in [TSH] has a sensitivity dependent on the values of [FT4].

6. During a condition of hyperthyroidism for a longer period we encounter a decreased value of model parameter S, resulting in a HP shift to the lower end of the [FT4] scale. When there is a prolonged period of hypothyroidism we find an increased value of model parameter S, resulting in a temporary shift to the higher end of the [FT4] axis.

7. During pregnancy we encounter an increased value of model parameter S resulting in a [FT4] shift of about 2 to 3 pmol/L to the higher end of the [FT4] scale.

As a last note, we have to state that the sensitivity of [TSH] values is an order of magnitude better than the sensitivity of [FT4]; however, the accuracy of both methods remains in the order of ± 5–10%.

For that reason, the detection of [TSH] in the range lower than 0.1 mU/L rapidly loses any significance. Because most relevant values of [FT4] are in the range of $5 < $ [FT4] $ < 50$ pmol/L, we have more confidence in the presented results.

References

[1] Goede, S. L., Leow, M. K., Smit, J. W., and Dietrich, J. W. (2014). A novel minimal mathematical model of the hypothalamus-pituitary-thyroid axis validated for individualized clinical applications. *Math. Biosci.* 249, 1–7. doi: 10.1016/j.mbs.2014.01.001

[2] Goede, S. L., and Leow, M. K. (2013). General error analysis in the relationship between free thyroxine and thyrotropin and its clinical relevance. *Comput. Math. Methods Med.* 2013, 831275. doi: 10.1155/2013/831275

[3] Leow, M. K., and Goede, S. L. (2014). The homeostatic set point of the hypothalamus-pituitary-thyroid axis–maximum curvature theory for personalized euthyroid targets, (2014) *Theor. Biol. Med. Model.* 11:35. doi: 10.1186/1742-4682-11-35

[4] Hörmann, R., Midgley, J. E., Larisch, R., and Dietrich, J. W. (2013). Is pituitary TSH an adequate measure of thyroid hormone-controlled homoeostasis during thyroxine treatment? *Eur. J. Endocrinol.* 168, 271–280.

[5] Midgley, J. E. M., Hörmann, R., Larisch, R., and Dietrich, J. W. (2013). Physiological states and functional relation between thyrotropin and free thyroxine in thyroid health and disease: in vivo and in silico

data suggest a hierarchical model. *J. Clin. Pathol.* 66, 335–342. doi: 10.1136/jclinpath-2012-201213

[6] Jonklaas, J., and Soldin, S. J. (2008). Tandem mass spectrometry as a novel tool for elucidating pituitary-thyroid relationships. *Thyroid* 18, 1303–1311.

[7] Roelfsema, F., Pereira, A. M., Veldhuis, J. D., Adriaanse, R., Endert, E., Fliers, E., et al. (2009). Thyrotropin Secretion Profiles Are Not Different in Men and Women. *J. Clin. Endocrinol. Metab.* 94, 3964–3967.

[8] Hennemann, G., Docter, R., Visser, T. J., Postema, P. T., and Krenning, E. P. (2004). Thyroxine plus low-dose, slow-release tri-iodothyronine replacement in hypothyroidism: proof of principle. *Thyroid* 14, 217–275.

[9] Schwarz, P., Madsen, J. C., Rasmussen, A. Q., Transbol, and I., Brown, E. M. (1998). Evidence for a role of intracellular stored parathyroid hormone in producing hysteresis of the PTH-calcium relationship in normal humans. *Clin. Endocrinol.* 48, 725–732.

[10] Finkbeiner, A. E. (1999). In vitro responses of detrusor smooth muscle to stretch and relaxation. *Scand. J. Urol. Nephrol. Suppl.* 201, 5–11.

[11] Doufas, A. G., Mastorakos, G. (2000). The hypothalamic-pituitary-thyroid axis and the female reproductive system. *Ann. N. Y. Acad. Sci.* 900, 65–76.

[12] Soldin, S. J., Soukhova, N., Janicic, N., Jonklaas, J., and Soldin, O. P. (2005). The measurement of free thyroxine by isotope dilution tandem mass spectrometry. *Clin. Chim. Acta.* 358, 113–118.

[13] Scarpelli, E. M., Gabbay, K. H., and Kochen, J. A. (1965). Lung surfactants, counterions, and hysteresis. *Science* 148, 1607–1609.

[14] Middelhoek, M. G. (1992). *The Identification of Analytical Device Models*. Ph.D thesis. Delft University Press, Delft.

[15] Soldin, O. P., and Soldin, S. J. (2011) Thyroid hormone testing by tandem mass spectrometry. *Clin. Bioichem.* 44, 89–94. doi: 10.1016/j.clinbiochem.2010.07.020

13

Model Identification, Validation, and Outlier Selection

"A model is a mathematical construct which, with the addition of certain verbal interpretations, describes observed phenomena. The justification of such a mathematical construct is solely and precisely that it is expected to work – that is, correctly to describe phenomena from a reasonably wide area."

–John von Neumann (1903–1957)

13.1 Introduction

13.1.1 A Brief History of Celestial Modeling as a Prelude to Model Validation

Celestial modeling is a brilliant way to start a discourse on the philosophy and science of model validation. It all began with keen observations of motion of celestial bodies. Astronomical observations dated as far back as the days of Stonehenge and the Babylonians (1,600 BC), making astronomy probably the very first scientific discipline. Ptolemy's geocentric model (140 AD) was based on the assumption that the Earth was positioned at the center of the universe. It was accepted as accurate within the measurement accuracies of the instruments of that time. In 1543, Copernicus, a Polish mathematician, proposed a heliocentric model in which the Sun lay at the center of our solar system, an idea originating from Aristarchus (~270 BC) though he was the first to promulgate and publish his model. Tycho Brahe, the Danish astronomer, formulated in 1560 a slightly different model in which the Earth remained in the center, but with Mars' orbit around the Sun, and the Sun–Mars composite orbit around the Earth. All these models provided relatively similar predictions of planetary positions.

Cosmology and astronomy serve as good examples of model validation based on previously observed and recorded data. The question remains: which model is better? This was the main question Johannes Kepler grappled with. Until then, the accepted paradigm was still the church's endorsed Aristotelian picture of the universe. Any unexplained anomaly (e.g., erratic retrograde motion of Mars) was thought to be part of the divine complexity of nature. In this era of entanglement of religious truths and realities, Kepler believed that a simpler model could be found to explain the observed phenomena. This moment came when he combined the observations of Tycho Brahe with his ideas of three-dimensional oppositions of Earth and Mars which led to his conclusion that Mars' orbit was not circular, but elliptic with the Sun in the center. In the new model, both Earth and Mars orbited the Sun elliptically which proved the superiority of this model above the epicycle idea of Ptolemy. Kepler also demonstrated that the orbiting speed of a planet is inversely proportional to the distance from the Sun.

With the invention and development of more accurate measurement equipment like the spyglass in 1608 by the Dutchman Hans Lippershey, the movement of planets could be observed with improved accuracy, and even more importantly, reveals objects that could not be observed with the naked eye. This idea of using glass optics to enhance distant objects inspired Galileo Galilei to several experiments and finally the construction of a telescope with a $30\times$ magnification power. Observations with this instrument revealed in great detail the satellites of the Jupiter and other phenomena in the planetary system that could not be previously observed. Because of these findings, Galileo deduced that the Sun should be the center of the solar system as we know it today.

While sharp observations propelled modeling accuracy, theoretical development was instrumental in explaining these observations and provided a robust scientific basis for the models. The work of Sir Isaac Newton resulted in the modern laws of mechanics and gravitation, published in the Philosophiae Naturalis Principia Mathematica in 1687. With these laws, the planetary orbits around the Sun could be predicted accurately. As well, Newton's law of universal gravitation could directly derive Kepler's laws of planetary motion. The validation of the central position of the Sun and the properties of the orbiting planets and moons in the solar system could be accomplished and established by increasingly accurate measurements, improved theoretical frameworks, and technology. However, the development of advanced celestial mechanical mathematics by Laplace and contemporary mathematicians in the 18th and 19th centuries indicated some doubts

regarding the Newtonian celestial theory. In 1846, the astronomer La Ver-
riere calculated that an unknown mass was influencing the orbit of Uranus.
This result led to the discovery of Neptune, and with similar observations,
measurements, and calculations, Pluto was also discovered. We have to keep
in mind that these achievements were possible chiefly due to the availability
of accurate and precise measurement equipment!

Thence, a "gold standard" of measurement and modeling endured until
better technologies and more advanced mathematical theories showed some
undeniable discrepancies between observations and Newtonian mechanics. A
very striking first positive confirmation of Einstein's general relativity was
demonstrated during measurements of the changes in relative positions of
stars behind the Sun due to gravitational lensing at a solar eclipse in 1919 by
Eddington and others. Also the theory of general relativity is now deemed the
more accurate model to establish the validity of cosmological measurements.
This process shows how increasing accuracy of technology when coupled
with the parallel development of new theoretical concepts is necessary and
synergistic with respect to modeling success and the capability of verifying
the theory with observations. In the following section, we will discuss the
validation and verification of biological models, and in particular, the HPT
model.

13.2 Model Identification

In Chapter 3, we discussed the method of observed data tables over the
range we had available in *experiment 2* where a curve-fitting method resulted
in the best function. In *experiment 1*, we could deduce the model from
the known physical properties of our object of investigation and found an
abstract mathematical expression. When we search for existing models of the
human HPT system, the collection is represented by mathematical models
and moreover physiological ones. None of these models have been validated
with the normal observable variables like [FT4] and [TSH].

The modeling of the HPT system can be divided in three main streams.

1. The clinical modeling based on statistical methods and looking for
   a non-dynamic, e.g., not time-dependent relationships and mostly the
   relationship between [TSH] and [FT4] [1–13].
2. The mathematical–biological modeling approaches of which the main
   focus is aimed at the dynamic aspects of hormone concentrations
   described with the wildest non-applicable but correct mathematical

derivations. None of these publications resulted in identifiable models nor could they be validated with real measured data [14–28].

3. Pharmacokinetic and pharmacodynamic modeling publications together with the developed theory were the only ones validated with real measured data [29].

Still, all modeling publications were aiming on a generalization of the phenomena and a generalized population model. This resulted in the fuzzy description and ordering of generalized averages not applicable for an individual.

In our modeling approach, we develop non-time-related parameterized relationships and dynamic time-dependent models dedicated to be applicable for an individual.

Most states of the HPT system are based on a dynamic equilibrium and can be described by means of parameterized static relationship.

A notable dynamic equilibrium modeling contribution was a mathematical treatise by Leow in 2007 which assumed that the relationship connecting [TSH] and [FT4] should take the form of a negative exponential function via the solution of a Bernoulli differential equation [21]. Together with many clinical and laboratory observations that the relationship between [TSH] and [FT4] had a possible LOG–LIN relationship, the publication of Suzuki et al. [9] presented a parameterized logarithmic expression which was inferred from incorrectly interpreted TSH–FT4 scatter plots of thousands of non-related euthyroid participants and without understanding the implications of this expression. The transformation of the Suzuki expression paved the way for our present modeling approach to arrive at a mathematically correct formal model, validated with clinical data that clinicians may apply for the individual management of thyroid patients [27].

Now we have arrived at the issue of model identification which implies that the available model parameters can be extracted from the measured data. In our model introduction in Chapter 6, is presented according to the following expression:

$$[TSH] = \frac{S}{\exp(\varphi[FT4])} \tag{13.1}$$

This mathematical abstraction describes the relationship between [FT4] and [TSH] by means of an exponential function and two model parameters. The proposed model describes a static relationship between the variables [FT4] and [TSH], and the model parameters S and φ can be calculated from measurements of which the curve has been derived accordingly:

$$\varphi = \left(\frac{1}{([FT4]_1 - [FT4]_2)}\right) \ln\left(\frac{[TSH]_2}{[TSH]_1}\right) \tag{13.2}$$

$$S = [TSH]_1 \exp(\varphi[FT4]_1) = [TSH]_2 \exp(\varphi[FT4]_2) \tag{13.3}$$

The parameters S and φ are calculated from the coordinates of the points

$$P_1 = ([FT4]_1, [TSH]_1) \tag{13.4}$$

and

$$P_2 = ([FT4]_2, [TSH]_2) \tag{13.5}$$

Two points are the theoretical minimum number of coordinates to define the values of S and φ which implies that there is one and only one exponential curve with these parameters determined by two points. With this property, we can easily verify the validity of our set of measurements, because the step of model identification has resulted in the model of Equation (13.1). In Figure 13.1, we present a clinical example of a set of thyroid function tests (TFTs) with a linear axis for [FT4] and [TSH]. Because the normal reference ranges of both [FT4] and [TSH] occur over the region of the exponential curve showing a pronounced bend, the location of the euthyroid set point

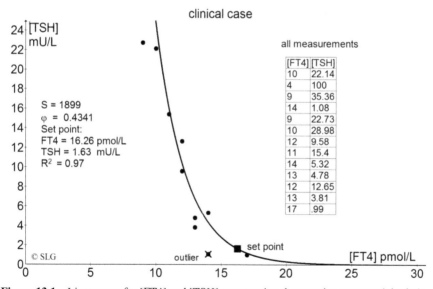

Figure 13.1 Linear axes for [FT4] and [TSH] representing the negative exponential relationship of [FT4] and [TSH]. In this plot, only one outlier is visible and the other one has a too high value of [TSH] for this [TSH] range.

is intuitively deduced to be located somewhere within this "knee" area. The visual detection of possible outliers is facilitated by the following transform of Equation (13.1):

$$\ln\left([TSH]\right) = \ln\left(\frac{S}{\exp(\varphi[FT4])}\right) \tag{13.6}$$

When we write

$$S = \exp(b), \tag{13.7}$$

we find

$$\ln\left([TSH]\right) = \ln\{\exp(b)\exp(-\varphi[FT4])\} = \ln\{\exp(b - \varphi[FT4])\}, \tag{13.8}$$

and with

$$\ln\left([TSH]\right) = \ln\{\exp(b - \varphi[FT4])\} = b - \varphi[FT4], \tag{13.9}$$

we transform (Equation 13.1) to a linear equation of [FT4] with a logarithmic scale for [TSH] as vertical axis represented by Equation (13.9), as was originally presented by Suzuki [9]. The transform of a non-linear characteristic to a linear one will be a great help with the validation of TFTs. This presentation also facilitates the visual inspection of the data plot and the identification of outliers. This is especially facilitated by the use of a log presentation for the [TSH] range and a linear presentation for the [FT4] axis where the final relationship becomes a straight line. Any deviation from the straight line, such as a non-exponential function, reflects either another model or possible faulty measurements. A set of [FT4]–[TSH] data points, measured according to calibrated clinical laboratory assays, will generally result in a representation according to the picture of Figure 13.1.

In Figure 13.2, the result of the transformed presentation of Figure 13.1 is depicted.

In the plot of Figure 13.2, the outliers are clearly identifiable as those significantly distant from the main straight line.

The development and validation of the model was published by Goede et al. [27]. Furthermore, the model parameters S and φ, calculated by means of an optimal curve fitting procedure, determine the total behavioral properties of the closed-loop HPT system. The validity of the operating range has been discussed in Chapter 12 where the physiological limitations determine the operating boundaries.

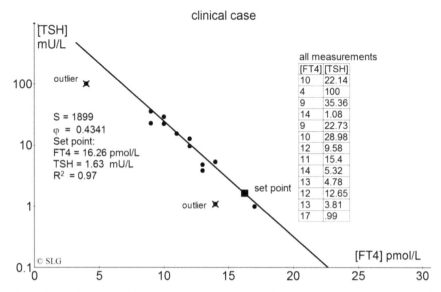

Figure 13.2 Log [TSH]–linear [FT4] presentation to identify outliers from the HP model.

As discussed, model selection depends on a pre-requisite accuracy over the validity domain of the model [28]. This enables us to define the criteria of outlier selection. An outlier is defined as a measurement result that cannot belong to the data set determined by the predefined accuracy sample space. This theory and belonging methods are described and discussed in detail by the dissertation of Middelhoek [28]. The model selection process has been finalized in Chapter 6 based on empirical gathered data from which a model conjecture and hypothesis resulted in the expression of (Equation 13.1) which now can be considered as a generalized law describing the relationship between $[FT4]$ and $[TSH]$. For every new patient treated for hypothyroidism with L-T4 toward the area of euthyroidism, every new stage of equilibrium provides a measured set of $[FT4]$ and $[TSH]$ that form the data points from which the model parameters S and φ can be calculated with the belonging curve fitting procedure. In the dissertation of Middelhoek [28], this fitting and outlier filtering method is described in the parameter extraction procedure called PARX and provides the calculated values of S and φ with the exclusion of calculated outliers. This PARX algorithm functions here as an outlier filter. Besides the curve-fitting and outlier filtering with PARX, there exist other acceptable alternatives for the curve-fitting procedure, but the outlier filtering is well suited for finding the best fit to the exponential model.

13.3 Examples of [TSH], [FT4], and [FT3] Outlier Identification

Besides model identification, we can find other types of outliers that can be identified from a series of measurements, as commonly done in trials over a defined period. Dynamic drug concentration behavior after drug ingestion can be analyzed with the methods described in the previous section when a physiologic model for its appearance and decay is defined. With regard to TFTs, [FT4] cannot rise in athyreotic patients who are not ingesting exogenous T4. The decay behavior of T4 is totally determined by its half-life, therefore a faster decay is very unlikely.

For all the hormones mentioned here, the decay from whatever position is limited to the half-life behavior. An unusually fast decay is circumspect and generally identifiable as an outlier.

In the following, we will discuss some examples of this type of outlier identification by means of a simple peak detection mechanism. This implies that the last highest level is remembered and stored, after which the half-life decay will begin until a next rise is detected and so on. The primary idea of such a peak detector is well known from radio receivers using amplitude modulation detection. In Figure 13.3, such a detector is depicted.

With the operating principle of the circuit depicted in Figure 13.3, we can detect and store the signal levels (i.e., hormone concentration levels equivalents), after which, without a level increase, the half-life decay will follow. In Figure 13.4, we depict a typical example of such a signal and detection behavior.

From Figure 13.4, we appreciate that the time-dependent signal of the hormone concentration U_1 starts in point (0,0) and rises to P1 at $t = 1$. After P1, we measure the value of P2 which is below the expected maximum decay

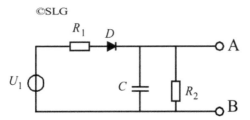

Figure 13.3   Diode top detector. U_1 represents the input signal of which the rising and peak levels will be stored in capacitor C. The resistor R_2 parallel to C represents together with C a decay time constant t equivalent to the half-life time constant of the measured hormone in question.

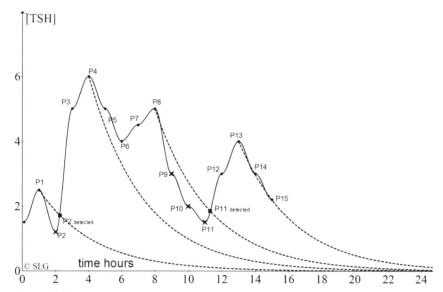

Figure 13.4 The peak detection procedure of signal U_1 with time sampling unit = 1 hour.

rate of the measured hormone concentration. Then the value of P2 is defined as an outlier. The minimum level of the output of the circuit of Figure 13.3 is then the value of P2$_{detected}$. From there, the rise to P3 and P4 will follow. After the peak of P4, the signal will decay again according to the half-life decay rate indicated by the dashed line. In this case, the decrease will remain above the half-life decay as indicated with the dashed line. All the measured values above the dashed line are supposed to be valid. At P8, we encounter a peak value again after which the half-life decay indicates that the values of P9, P10, and P11 are below the dashed line and therefore are outliers. From P11$_{detected}$, we see a rise via P12 to the next peak P13. From P13, the half-life decay dashed line indicates that the measured values of P14 and P15 are valid.

13.4 Discussion

When we have to find a model of which only a set of measurements is available, it is virtually impossible to construct a theory that supports the measurements. In endocrine systems, it is rather difficult to perform measurements with an accuracy normally encountered in isolated physical processes. For example, the accuracy of lengths' measurements can literally reach atomic scales, whereas hormone concentrations in clinical practice are at best currently in the order of 1–5%. Hormone concentrations are also part of dynamic processes of a living being like the circadian variations over 24 h. This forms an extra layer of pre-analytical complicating factor in the performance of a measurement procedure. Furthermore, the conditions under which a test is done can present different and sometimes erratic results indicated as test noise in addition to intra-assay and inter-assay variability due to analytical noise. The effects of measurement noise can be reduced by the identification of the underlying model to help in the process of data filtering and removing the outliers, and come to the best fitting result.

References

Clinical [FT4] – [TSH] models

[1] Reichlin, S., and Utiger, R. D. (1967). *Regulation of the Pituitary-Thyroid Axis in Man: Relationship of TSH Concentration to Concentration of Free and Total Thyroxine in Plasma.* Available at: JCEM endojournals.org

[2] Ercan-Fang, S., Schwartz, H. L., Mariash, C. N., and Oppenheimer, J. H. (2000). Quantitative assessment of pituitary resistance to thyroid hormone from plots of the logarithm of thyrotropin versus serum free thyroxine index. *J. Clin. Endocrinol. Metab.* 85:6.

[3] Fisher, D. A., Schoen, E. J., Franchis, S., Mandel, H., Nelson, J. C., Carlton, E. I., et al. (2000). the hypothalamic-pituitary-thyroid negative feedback control axis in children with treated congenital hypothyroidism. *J. Clin. Endocrinol. Metab.* 85:8.

[4] Falaschi, P., Martocchia, A., Proietti, A., D'Urso, R., Gargano, S, Culasso, F., et al. (2004). The hypothalamic-pituitary-thyroid axis in subjects with subclinical thyroid diseases: the impact of the negative feedback mechanism. *Neuroendocrinol. Lett.* 25.

[5] Jonklaas, J., and Soldin, S. J. (2008). Tandem mass spectrometry as a novel tool for elucidating pituitary–thyroid relationships. *THYROID* 18. doi: 10.1089=thy.2008.0155

[6] Jostel, A., David, W., Ryder, J., and Shalet, S. M. (2009). The use of thyroid function tests in the diagnosis of hypopituitarism: definition and evaluation of the TSH Index, 2009. *Clin. Endocrinol.* 71, 529–534. doi: 10.1111/j.1365-2265.2009.03534.x

[7] Benhadi, N., Fliers, E., Visser, T. J., Reitsma, J. B., and Wiersinga, W. M. (2010). Pilot study on the assessment of the setpoint of the hypothalamus–pituitary–thyroid axis in healthy volunteers. *European Journal of Endocrinology* 162, 323–329. doi: 10.1530/EJE-09-0655

[8] van Deventer, H. E., Mendu, D. R., Remaley, A. T., and Soldin, S. J. (2011). Inverse log-linear relationship between thyroid-stimulating hormone and free thyroxine measured by direct analog immunoassay and tandem mass spectrometry. *Clin. Chem.* 57, 122–127.

[9] Suzuki, S., Nishio, S., Takeda, T., and Komatsu, M. (2012). Gender-specific regulation of response to thyroid hormone in aging. *Thyroid Res.* 5:1.

[10] Hadlow, N. C., Rothacker, K. M., Wardrop, R., Brown, S. J., Lim, E. M., and Walsh, J. P. (2013). The relationship between TSH and free T4 in a large population is complex and nonlinear and differs by age and sex. *J. Clin. Endocrinol. Metab.* 98, 2936–2943. doi: 10.1210/jc. 2012-4223

[11] De Grande, L. A. C., Van Uytfanghe, K., and Thienpont, L. M. A. (2015). Fresh look at the relationship between TSH and free thyroxine in cross-sectional data. *Eur. Thyroid J.* 4, 69–70. doi: 10.1159/000369796

[12] Rothacker, K. M., Brown, S. J., Hadlow, N. C., Wardrop, R., and Walsh, J. P. (2016). Reconciling the log-linear and non-log-linear nature of the TSH-free T4 relationship: intra-individual analysis of a large population. *J Clin Endocrinol Metab.* doi: 10.1210/jc.2015-4011

[13] Fitzgerald, S. P., and Bean, N. G. (2016). The relationship between population T4/TSH set point data and T4/TSH physiology. *J. Thyroid Res.* 2016:6351473. doi: 10.1155/2016/6351473

Mathematical [FT4] – [TSH] models

[14] Danziger, L., and Elmergreen, G. L. (1956). The thyhroid – pituitary homeostatic mechanism. *Bull. Math. Biophys.* 18.

[15] DiStefano, J. J., and Stear, E. B. (1968). Nueroendocrine control of thyroid secretion in living systems. *Bull. Math. Biophys.* 30.

[16] DiStefano, J. J. (1969). A model of the normal thyroid hormone glandular secretion mechanism. *J. Theor. Biol.* 22, 412–417.

[17] DiStefano, J. J., Wilson, K. C., Jang, M., and Mak, P. H. (1975). Identification of the dynamics of thyroid hormone metabolism. *Automatica* 11, 149–159.

[18] Liu, Y., Liu, B., Xie, J., Liu, Y. X. (1994). A new mathematical model of hypothalamo-pituitary-thyroid axis. *Math. Comut. Model.* 19, 81–90.

[19] Degon, M., Chait, Y., Hollot, C. V., Chipkin, S., and Zoeller, T. (2005). "A Quantitative Model of the Human Thyroid: Development and Observations," in *Proceedings of the 2005 American Control Conference*, June 8–10, 2005, Portland, OR, USA.

[20] Mukhopadhyay, B., and Bhattacharyya, R. (2006). A mathematical model describing the thyroid-pituitary axis with time delays inhormone transportation. *Appl. Math.* 51, 549–564.

[21] Leow, M. K. (2007). A mathematical model of pituitary–thyroid interaction to provide an insight into the nature of the thyrotropin–thyroid hormone relationship. *J. Theor. Biol.* 248, 275–287.

[22] Eisenberg, M., Samuels, M., and DiStefano, J. J. (2008). Extensions, validation, and clinical applications of a feedback control system simulator of the hypothalamo-pituitary-thyroid axis THYROID. 18. doi: 10.1089=thy.2007.0388

[23] Eisenberg, M. C., Santini, F., Marsili, A., Pinchera, A., and DiStefano, J. J. (2010). TSH regulation dynamics in central and extreme primary hypothyroidism. *Thyroid* 20, 11. doi: 10.1089/thy.2009.0349

[24] Seker, O. (2012). Modeling the dynamics of thyroid hormones and related disorders. Master thesis, Boğaziçi University, Istanbul.

[25] Pandiyan, B, Merrill, S. J., and Benvenga, S. (2013). A patient-specific model of the negative-feedback control of the hypothalamus–pituitary–thyroid (HPT) axis in autoimmune (Hashimoto's) thyroiditis. *Math. Med. Biol.* doi: 10.1093/imammb/dqt005

[26] Goede, S. L., and Leow, M. K. (2013). General error analysis in the relationship between free thyroxine and thyrotropin and its clinical relevance. *Comput. Math. Methods Med.* doi: 10.1155/2013/831275

[27] Goede, S. L., Leow, M. K., Smit, J. W. A., and Dietrich, J. W. (2014). A novel minimal mathematical model of the hypothalamus–pituitary–thyroid axis validated for individualized clinical applications. *Mat Biosci.* doi: 10.1016/j.mbs.2014.01.001

[28] Middelhoek, M. G. (1992). *The Identification of Analytical Device Models*. Ph.D. thesis, Delft University of Technology, Delft.

Pharmacokinetic models

[29] Bauer, L. A. (2014). *Applied Clinical Pharmacokinetics*, 3rd Edn. New York, NY: McGraw-Hill.

14

Half-Life and Plasma Appearance Dynamics of T3 and T4

"Mathematics is much more than a language for dealing with the physical world. It is a source of models and abstractions which will enable us to obtain amazing new insights into the way in which nature operates."

–Melvin Schwartz (1932–2006)

14.1 Introduction

The absorption, distribution, appearance, storage, metabolism, excretion, and secretion of substances by the body fall under the general principles of pharmacokinetics (PK) whereas the effects of the substances on the body are described in terms of pharmacodynamics (PD) [1]. In this chapter about the analysis of PK and PD properties of the hypothalamus–pituitary–thyroid (HPT) system related to L-T3 and L-T4, we work under the assumption that the medication with synthetic thyroid hormones is a replacement for the normal thyroid production and is aimed to reach a homeostasis near the HPT axis set point [2]. Because every person has individual physiological hypothalamus–pituitary (HP) parameter values which can substantially differ from others, it is impossible to provide a general fixed model of the behavior for dosing of medications, because the PK and PD parameters of a person are fundamentally unique as is demonstrated in a recent paper [3]. This aligns to the prerogative of the new era of personalized medicine that individual metabolic differences need to be taken into account insofar as medication dosing is concerned as there can be significant projected impact with over or under dosing.

In this chapter, we will introduce the modeling principles using an electrical network theory that incorporates PK/PD. The results could be easily applied in practical situations by means of open source available electrical

187

network simulators [4]. Such modeling allows the determination of bio-equivalence and analyses of differences between various generic brands of L-T4. The potency differences can be evaluated using healthy reference persons to test the appearance, C_{max}, T_{max}, and AUC (Area Under Curve) of one brand at period 1 and repeated using a comparator brand at a later period. The differences per individual can then be demonstrated.

14.2 Half-Life ($t_{1/2}$) of T4 and Calculation of the Decay Time Constant

Every molecule in the human body that is subjected to metabolism or other relevant biological processes exhibits a half-life defined as the time for its concentration to decrease by 50%. The notion of half-life is closely associated with a negative exponential decay process. In general, a single exponential, single distribution compartment first-order decay processes may be modeled as:

$$S = K \exp(-\lambda t) \tag{14.1}$$

This type of decay process is primarily related to the leaking fluid container where S is equivalent with the fluid level and λ is related to the leakage opening in the bottom.

The remaining amount of the substance S is a function of time t. The process starts at $t = 0$, which implies the initial amount $S = K$. After some time t, we reach $S = 0.5K$. This occurs when:

$$K \exp(-\lambda t) = 0.5K, \tag{14.2}$$

or

$$\exp(-\lambda t) = 0.5, \tag{14.3}$$

resulting in

$$-\lambda t = \ln(0.5) = {}^{e}\log(0.5) = -0.693 \tag{14.4}$$

Then, the half-life value for t follows:

$$t = \frac{0.693}{\lambda}, \tag{14.5}$$

which implies that the half-life time is dependent on λ.

The half-life of a process is defined as the time necessary for a substance to decline to the half of the original amount. This time, $t_{1/2}$, is a constant across a negative exponential decay process behavior of first-order kinetics.

$$A(t) = A_0 2^{(-t/t1/2)} \tag{14.6}$$

This type of decay is typically observed for many processes in nature, particularly in the area of physiology.

Instead of S, the original amount is here defined as A_0.

Over a short period of time,

$$t_1 - t_2 = \Delta t, \tag{14.7}$$

we see a decay from amount A_1 to amount A_2 resulting in

$$A_1 - A_2 = \Delta A \tag{14.8}$$

The speed of this decay process is the amount of decay per unit of time resulting in

$$\frac{A_1 - A_2}{t_1 - t_2} = -K_1 A \tag{14.9}$$

or

$$\frac{\Delta A}{\Delta t} = -K_1 A \tag{14.10}$$

For infinitesimally small differences of ΔA and Δt, we can write $\Delta A = dA$ and $\Delta t = dt$, from which we obtain the differential equation:

$$\frac{dA}{dt} = -K_1 A \tag{14.11}$$

or

$$\frac{dA}{A} = -K_1 dt \tag{14.12}$$

$$\int_0^t \frac{dA}{A} = -\int_0^t K_1 dt \tag{14.13}$$

$$\ln(A_t) - \ln(A_0) = -K_1 t, \tag{14.14}$$

or

$$\ln\left(\frac{A_t}{A_0}\right) = -K_1 t, \tag{14.15}$$

then

$$A_t = A_0 \exp(-K_1 t) \tag{14.16}$$

When we write for:

$$K_1 = \frac{1}{\tau}, \tag{14.17}$$

we can state that the amount or concentration will be decreased with a factor $1/e$ in a time τ.

The factor τ is called the first-order time constant. Because half-life time is related to the number 2 or $^1\!/_2$, we can rewrite A_t as:

$$A_t = A_0 \exp(-K_1 t) = A_0 2^{(-(t/\tau)\ln 2)}, \qquad (14.18)$$

and because

$$\tau \ln(2) = t_{1/2}, \qquad (14.19)$$

we find

$$A_0 2^{(-(t/\tau)\ln 2)} = A_0 2^{(-t/t1/2)}. \qquad (14.20)$$

With $1/\ln(2) = 1.4427$, we find for the general time constant for the half-time

$$\tau = \frac{t_{1/2}}{\ln(2)} = 1.4427 t_{1/2} \qquad (14.21)$$

According to the natural negative exponential decay process, circulating thyroid hormone (T3 or T4) levels may be formulated as:

$$A_t = A_0 \exp(-t/\tau) \qquad (14.22)$$

A_0 represents the value of the accumulated T4 or [FT4] and A_t represents the value as a function of time.

14.2.1 Half-Life Modeling with Electrical Networks

The use of electrical network equivalents was successfully introduced by the work of Hodgkin et al. [5, 6]. They proved that electrical network elements perfectly reflect the behavior of characterized physiologic functions. The electrical network analysis methods can be studied in detail in the work of Schenkman [7]. In the following, we use straightforward elementary network theory to calculate the various voltages and currents based on Ohm's law and the elementary principles from electrodynamics.

The removal and/or metabolism of a substance from the blood plasma defined according to the previous part can also be described with the notion of a capacitor discharge by a resistor. This discharge process is depicted via the electrical network in Figure 14.1.

We use Ohm's law to analyze the static circuit

$$U = I.R \ \text{ or } \ I = \frac{U}{R}, \qquad (14.23)$$

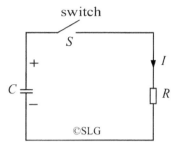

Figure 14.1   Network representation capacitor discharge.

whereby U represents the voltage across the resistor R and I represents the value of the current in the circuit. The general equation for the total charge Q is defined as

$$Q = C.U_C \tag{14.24}$$

$$I = \frac{dQ}{dt} = -C\frac{dU_C}{dt} \tag{14.25}$$

The minus sign in the equations follows from the fact that the capacitor is being discharged.

Equation (14.25) represents the relationship between the capacitor current I and belonging capacitor voltage U_C. From Figure 14.1, we read the value of I when the switch is closed:

$$I = \frac{U_C}{R} \tag{14.26}$$

When we combine Equations (14.25) and (14.26), we find

$$\frac{U_C}{R} = C\frac{-dU_C}{dt}, \tag{14.27}$$

resulting in

$$U_c = -RC\frac{dU_C}{dt}, \tag{14.28}$$

or

$$\frac{dU_C}{U_C} = \frac{-1}{RC}dt \tag{14.29}$$

Equation (14.29) is a first-order linear differential equation that can be solved by integration of both sides.

Thus, we find:

$$\int \frac{dU_C}{U_C} = \int \frac{-1}{RC} dt,$$

(14.30)

resulting in

$$\ln(U_C) = \frac{-t}{RC} + K$$

(14.31)

Because the product RC has the dimension of time, we can define the time constant

$$\tau = RC$$

(14.32)

Equation (14.32) can then be written as:

$$U_C = \exp\left(\frac{-t}{\tau} + K\right) = \exp(K).\exp\left(\frac{-t}{\tau}\right)$$

(14.33)

From the boundary conditions, we can calculate the value of exp(K) = U.
For the time-dependent form of the capacitor voltage, $U_C(t)$, we find:

$$U_C(t) = U \exp\left(\frac{-t}{RC}\right)$$

(14.34)

$$\tau = \frac{t_{1/2}}{\ln(2)} = 1.4427 t_{1/2}$$

(14.35)

or

$$t_{1/2} = RC \ln(2)$$

(14.36)

With these results, the following PK/PD will be discussed based on an electrical network theory.

14.3 Drug Input and Output Signals

Let us follow the drug concentration variables as a function of time from an oral L-T3 tablet bolus of D in the blood compartment. The human body is a system in which drugs can be absorbed, transported, stored, metabolized, and excreted. The gastrointestinal tract (GIT) acts as a processing system governing the entry of drugs in an individually determined, time-dependent fashion to the blood compartment that serves as a generalized transport and storage system. Upon consumption of a bolus D, the drug profiles in the blood plasma can vary depending on the dissolution, pH, absorption rate, hydrophobicity, protein binding, partitioning in tissue compartments, storage, utilization, and excretion.

In the following, the drug concentrations as a function of time will be regarded as signals.

This explanation is important because the discussed mechanism stands as a generalized template for all other substances we take in as food, drinks, and also drugs. The effect of the intake is equivalent to a certain bolus of material that builds during the process of eating and drinking. The drug intake process and digestion result in a linear output to the blood compartment. The result from the gut to the blood compartment is a rather smooth continuous rising level of the digested intake. This continuous rising process ends the moment the metabolic conversion from the GIT is finalized because all the ingested material has been used. The GIT thus exhibits an integration effect on the bolus. During the rise of the output from the GIT, a second integration process is performed by the blood compartment which operates as a storage capacitor with a defined capacitance C and responds on a time-dependent driving function $f(t)_D$ in units of ng/dL/sec in series with a transport resistor.

The signal path is depicted in Figure 14.2.

The Laplace transform is defined as

$$F(s) = \int_0^\infty f(t)\exp(-st)dt. \tag{14.37}$$

This integral transform can be interpreted as a function transformer.

In Figure 14.3, we depict the calculation diagram along which complicated differential equations which can be transformed from the time domain to the s domain as normal algebraic equations. After solving the equations in the s domain, we retransform the algebraic equations back to the time domain with the inverse Laplace transform according to

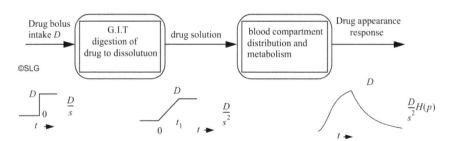

Figure 14.2   Signal processing compartment, tracing the path of the drug from the digestion of a single bolus D in the gastrointestinal tract (GIT) to its trajectory in the circulation according to the principles of drug PK. Each time-function pictogram and its associated Laplace transformed signal are illustrated at every stage.

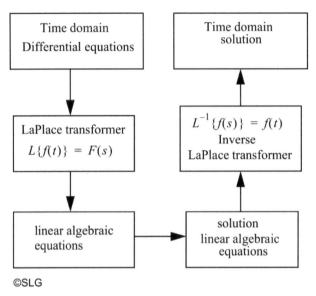

©SLG

Figure 14.3   Calculation diagram for the application of Laplace transforms.

$$f(t) = L^{-1}\{F(s)\} = \int_{\alpha-j\infty}^{\alpha+j\infty} \exp(st)F(s)ds \qquad (14.38)$$

From Figure 14.2, we appreciate that the drug bolus D is approximated by a step function

$$D\varepsilon(t), \qquad (14.39)$$

resulting in a Laplace transformed presentation,

$$\frac{D}{s}, \qquad (14.40)$$

where

$$s = j\omega, \qquad (14.41)$$

with $j^2 = -1$ and ω represents the frequency, and D represents the maximum amount of the dose. The GIT operates as a one-directional integrating converter from which the $f(t)_D$ signal, as a driving function of time, is represented by a ramp function,

$$f(t)_D = \int_0^{t1} D\varepsilon(\tau)d\tau = \frac{D}{t_1}t, \qquad (14.42)$$

with the Laplace-transformed presentation

$$\frac{D}{s^2} \tag{14.43}$$

This generalized signal is the input of the blood compartment capacitor via an input resistance, which is represented by the physiological process of time necessary to transport a defined amount of molecules from one location to another. For the graphical presentations, we use the open source mathematical graphical analysis tool: Graph 4.4. The general driving function $f(t)_D$ is depicted in Figure 14.3 as a straight line from the points (0,0) to (t_1, D) and shows the maximum T3 concentration level D after t_1 hours, corresponding with the maximum amount of equivalent T3 bolus.

The signal is presented without the level of the baseline [T3], as is shown later in the appearance and decay signal presentations and with interval operating constraints

$$0 < t < t_1, \tag{14.44}$$

for the linear rising part.

The maximum value D of the input signal is maintained over the interval of

$$t_1 < t < t_2 \tag{14.45}$$

Figure 14.4 shows the signal output of the GIT which is approximated as a linearly increasing signal during a defined period t_1. This signal is consistent

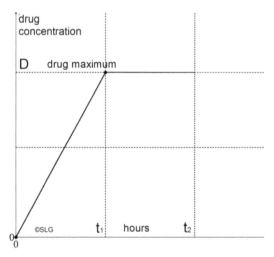

Figure 14.4 $f(t)_D$ output from the gastrointestinal process with a maximum at max drug concentration = D over the interval $0 < t < t_1$ and sustains this value at D during $0 < t < t_2$.

with the way the drug bolus is digested, followed by gradual absorption from the GIT surface, and transferred across into the circulation during bolus processing. The drug concentration rise from the GIT behaves approximately as an integration of the bolus step function of Equation (14.39). The second part of the driving function is described over the interval $t_1 < t < t_2$, where the maximum value is maintained until t_2.

In Figure 14.5, the general structure of the drug appearance and decay is presented with the time functions related to the designated time intervals.

At time $t = t_1$, the maximum drug concentration level D has been reached, the integration operation on $f_1(t)$ will end, and the appearance function continues with the driving function as:

$$f(t)_D = D\varepsilon(t) = \frac{D}{t_1}t, \tag{14.46}$$

over the interval $t_1 < t < t_2$.

The resulting differentiation operation on $f_1(t)$ is written as:

$$f_2(t) = f_1(t_1) + \frac{df_1(t - t_1)}{dt}(t_1 < t < t_2) \tag{14.47}$$

because at $t = t_2$, the resolution process of the liothyronine bolus has ended. The level of D is no longer sustained, the decay process will begin, and

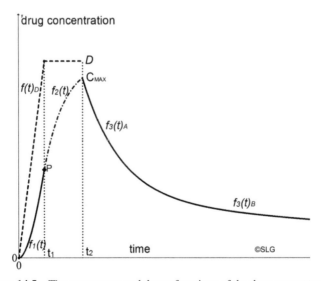

Figure 14.5   The appearance and decay functions of the drug concentration.

the rise of $f_2(t)$ is limited to this point in time. The decay function of the distribution phase $f_3(t)$ is indicated with a solid line as the output signal of the blood compartment. The PK decay profile is subdivided into a distinct distribution phase occurring directly after C_{max} as $f_3(t)_A$ for about 10 h, followed by an elimination phase $f_3(t)_B$ consistent with a bi-exponential decay function. The specific details will be discussed in the following section.

14.3.1 Network Model

From measurements of drug appearance, it becomes intuitively obvious that every stage of this process chain has an integrating character. The drug signal appearance and utilization can therefore be subjected to the fundamental laws of physics of electricity and hence be modeled as an integrating electrical circuit from which the input and output signals can be verified either by conventional calculation, or with possibilities to investigate variations in real time by means of a circuit simulator.

For the derivation of the first part of the appearance function, we will use this first-order integrator model. The modeling approach of the bi-exponential decay will be discussed later with an extra integrating section. The analysis of an electrical network facilitates the modeling and simulation with open available tools. The endocrine dynamics and associated signal behavior can now be described and analyzed where the time-domain functions are transformed to the $j\omega$ domain or complex frequency $s = j\omega$, using the Laplace transform.

With this transform, we can calculate the currents and voltages in an AC network without the use of differential equations which would expand in the time domain to unworkable proportions. Therefore, we use the Laplace transform for all network elements resulting in simple algebraic equations that can easily be solved in this transformed domain. The analyzed results and solutions can subsequently be inversely transformed back to the time domain. After this mathematical procedure, the analysis will be continued with time functions.

In Figure 14.6, the electrical circuit represents the physiological model equivalent of the blood compartment. When we follow the drug concentration as a result of absorption from the GIT, we can define this as equivalent to an electrical voltage source U. Diode D prevents backflow because the metabolic conversion processes resulting from the drug bolus are essentially irreversible. Resistor R_1 represents the resistance from the GIT to the blood compartment. Capacitor C represents the volume of this blood compartment while R_2 stands for all integrated peripheral drug level-dependent metabolism. A high value of R_2 indicates a low discharge current and thus

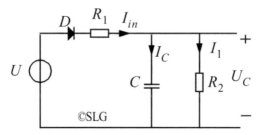

Figure 14.6   Elementary electrical circuit equivalent of the blood compartment with a driving source U. This causes a current I_{IN} during the resolution process via the tissue related resistor R_1 in a charge for blood compartment capacitor C. The diode D represents the unidirectional flow of the resolved amount of T3. The metabolism-related resistor R_2 represents the discharge of the capacitor and the belonging decay process. The combination C and R_2 represents a single-compartment PK decay process.

a slow decay of the drug. A small value of R_2 has the opposite effect. In a single-compartment model, the first-order decay is manifested as a single negative exponential coefficient which represents the half-life behavior and is here represented by the time constant of the combination R_2 and C.

Generally, resistance is defined as the time it takes to pass a defined amount of material, from one place to another, with the dimension seconds/liter equivalent to the electrical element R in ohms. The drug charging current for the capacitor C is defined as I_{IN} and is the amount of material passed per second with the dimension ng/sec equivalent to the current strength unit in amperes. We use the following equations in analogy from electrodynamics theory.

The amount of the liothyronine l in the circulation following oral absorption:

$$l = C.U \text{ ng} \qquad (14.48)$$

The transported amount of drug per unit of time

$$I = \frac{dl}{dt} \text{ ng/sec} \qquad (14.49)$$

This results in a relationship between the capacitor current I_C and capacitor voltage

$$I = I_C = C\frac{dU_C}{dt} \text{ ng/sec} \qquad (14.50)$$

The drug concentration level is equivalent to a voltage value U ng/dL. The expression of Equation (14.50) forms the basis of all related differential equations describing the dynamics of the drug concentration behavior. The related differential equations will be presented as Laplace transformed expressions in the $j\omega$, or s domain.

With

$$\frac{d}{dt}U_C = sU_C,\tag{14.51}$$

the Laplace transform of the capacitor impedance results in

$$Z_C = \frac{U_C}{I_C} = \frac{1}{sC},\tag{14.52}$$

with s as the complex frequency $s = j\omega$.

The resistor is defined as

$$R = \frac{U}{I} \ \text{sec/L}\tag{14.53}$$

We will define the capacitor value proportionally to the individual body weight according to $C = 4.5$ ml/kg representing the volume in liters of the blood compartment. Based on the capacitor value, we can derive the values for the belonging resistors R_1 and R_2. With these relationships, we translate the physiological parameters to electrical network elements and vice versa. The electrical network elements R_1, R_2, and C have fixed unique values for any given individual and mainly represent the model parameters for the appearance phase. In the time domain, U represents the input signal in the form

$$U(t) = \frac{D}{t_1}t,\tag{14.54}$$

for the time interval $0 < t < t_1$ as shown in Figure 14.4. Because we are interested only in the dynamics of the signal behavior, we can write the transfer of the electrical network of Figure 14.5 as the Laplace-transformed time function.

$$\frac{U_C}{U} = \left(\frac{R_2}{R_1 + R_2}\right)\frac{1}{(s\tau + 1)}\tag{14.55}$$

Appearance time constant $\tau = \tau_1$ is denoted by

$$\tau_1 = \left(\frac{R_1 R_2}{R_1 + R_2}\right)C\tag{14.56}$$

The decay time constant τ_D for a single-compartment decay

$$\tau_D = R_2 C\tag{14.57}$$

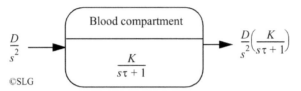

Figure 14.7   Laplace-transformed signal transfer.

Going back to the model depicted in Figure 14.2, we can derive the appearance characteristic for a single bolus of 50 μg liothyronine. The T3 output in the blood compartment after oral ingestion of liothyronine is presented as a linear function of time.

This can be described according to Figure 14.7.

We use the following basic Laplace transform for the output of the GIT. Under the condition that $R_2 \gg R_1$, which is valid for most cases, we can approximate

$$K \approx \frac{R_2}{R_1 + R_2} \approx 1 \tag{14.58}$$

For the Laplace transformed signal U, we find:

$$U(p) = \frac{D}{s^2} \tag{14.59}$$

The output signal $U_C(s)$ represents the measured values of the T3 concentrations in the blood compartment.

$$U_C(s) = \frac{DK}{s^2(p\tau + 1)} = \frac{DK}{s^2} - \frac{DK\tau}{s} + \frac{DK(\tau)^2}{s\tau + 1} = \frac{DK}{s^2} - \frac{DK\tau}{s} + \frac{DK\tau}{s + 1/\tau}$$

$$\tag{14.60}$$

When $U_C(s)$ as resolved above into partial fractions are inversely transformed to the time domain and we then apply the substitution:

$$a = \frac{1}{\tau}, \tag{14.61}$$

we obtain the piecewise continuous function $U_C(t)$ which is composed of two partial functions.

For $K \approx 1$, we have the first part of the drug appearance with an output range defined by

$$U_{C1}(t) = f_1(t) = \frac{D}{t_1}[t - \tau\{1 - \exp(-at)\}] \quad \text{for } 0 < t < t_1, \tag{14.62}$$

as indicated in Figure 14.3.

From $t = t_1$, the driving function $f(t)_D$ will not rise anymore and remains at the end level $U_L = A$ until $t = t_2$ causing the appearance function to continue as

$$U_{C2}(t) = f_2(t) = U_{C1}(t_1) + \frac{df_1(t - t_1)}{dt}, \tag{14.63}$$

resulting in

$$U_{C2}(t) = U_{C1}(t_1) + \{D - U_{C1}(t_1)\} \{1 - \exp(-a(t - t_1))\}, \tag{14.64}$$

over the interval $t_1 < t < t_2$, and represents the saturating part of the appearance response.

The function $f_2(t)$ is truncated at $t = t_2$ because the bolus liothyronine has been completely absorbed, preventing $f_2(t)$ to reach the saturation level A.

In order to connect $U_{C1}(t)$ seamlessly at $t = t_1$ to $U_{C2}(t)$, the following condition has to be met:

$$\frac{d\{U_{C1}(t)\}}{dt} = \frac{D}{t_1}[1 - \exp(-at_1)] \quad \text{at } t = t_1, \tag{14.65}$$

and equals to the first derivative of

$$\frac{d\{U_{C2}(t)\}}{dt} = \frac{\{D - U_{C1}(t_1)\}}{\tau} \quad \text{at } t = t_1 \tag{14.66}$$

This results in a non-symmetrical sigmoidal appearance response curve in which the beginning trajectory from $0 < t < t_1$ has a different curvature compared to the second part with $t_1 < t < t_2$ and is in general asymmetrical. However, both parts of the appearance functions are defined by the same time constant τ. The time signal $U_C(t)$ is determined by three parameters, A, K, and τ.

14.4 Decay Behavior after Finalization of Drug Appearance

The modeling of the redistribution of drug uptake is important because the definition of half-life is not valid for a multi-compartmental distribution phenomenon. The half-life definition is based on a single-compartment drug distribution behavior where the half-life effect is a result of the internal metabolic processes. We designate this as linear PK, which means that the elimination behaves according to a first-order negative exponential decay, as is modeled with the electrical circuit equivalent of Figure 14.8.

$$U_0 = U \exp\left(-\frac{t}{R_2 C_1}\right) \tag{14.67}$$

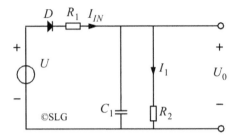

Figure 14.8 First-order elimination of drug content represented by U_0.

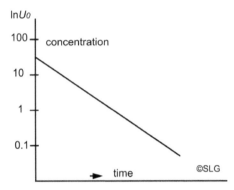

Figure 14.9 First-order decay of $\ln(U_0)$.

When we depict this behavior with a logarithmic scale for the concentration and a linear scale for time, we can depict the behavior of U_0 according to Figure 14.9.

A two-compartment kinetic behavior is well described in literature [1] and results in a parameterized bi-exponential function. After the moment C_{max} has been reached, a typical two-phase decay profile can be observed. This type of decay is modeled with two compartments [5] and can be described with the electrical network model of Figure 14.10.

The effect of the two-compartment model can be similarly depicted as a time function as was done in Figure 14.9. In Figure 14.11, we see a decay divided in two sections as a function of time. The first one which is relatively steep is indicated as the α distribution phase (plasma) defined by time constant $\tau_\alpha = C_1 R_2$ whereas the following phase indicated as elimination phase or β phase (tissue) shows a less steep decay defined by $\tau_\beta = C_2 R_3$.

With the configuration of Figure 14.10, we can derive the signal transfer from input signal U to the capacitor output voltage U_0 which is the summation of two different decay processes. The rapid α decay is represented by the

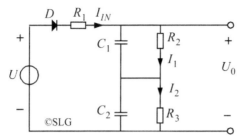

Figure 14.10 Electrical network model of a two-compartment decay model. The single-compartment model consisting of C_1 and R_2 is extended by a second compartment model consisting of C_2 and R_3.

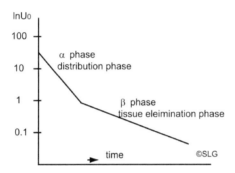

Figure 14.11 Timing diagram of the two-compartment model.

time constant formed by R_2 and C_1 as

$$\alpha = 1/\tau_\alpha, \tag{14.68}$$

where

$$\tau_\alpha = R_2 C_1 \tag{14.69}$$

The second or β phase of the decay is defined by

$$\beta = 1/\tau_\beta, \tag{14.70}$$

where

$$\tau_\beta = R_3 C_2 \tag{14.71}$$

The serial impedance of the load circuit formed by C_1, R_2, and R_3 is expressed as

$$Z_L = \frac{R_2}{sR_2C_1 + 1} + \frac{R_3}{sR_3C_2 + 1}, \tag{14.72}$$

from which we can find the inverse Laplace transform.

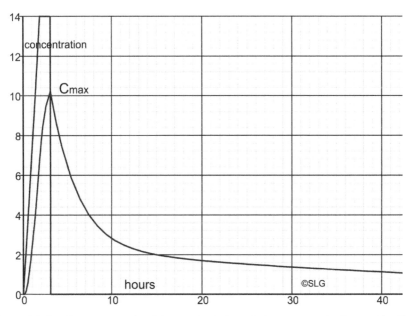

Figure 14.12   Simulated result of the configuration of Figure 14.8 with $R_1 = 1k$, $C_1 = 1000\ \mu F$, $R_2 = 3k$, $R_3 = 10k$, and $C_2 = 5000\ \mu F$. The vertical scale of [FT3] is in ng/dL, while time in the horizontal axis is in hours. This simulation is performed in the "seconds" time scale for convenience using the electrical circuit simulator and the time scale can be transformed to hours.

This result can be synthesized by means of a simulation of the circuit configuration of Figure 14.10 which is depicted in Figure 14.12.

The initial part (the alpha phase) is the rapid decrease with a relatively low time constant and is indicated by the distribution phase where the drug distributes from the circulation compartment across to other tissue compartments. The second part (beta phase) appearing after the distribution phase is a redistribution of the drug concentration from other tissue compartments back into the blood and to general drug metabolism and excretion/secretion which is associated with the normal half-life of the drug. This type of two-compartmental decay is observed for [T3] and [FT3] decay profiles.

The data points could be fitted with the following parameterized biexponential function.

$$f_3(t) = A\exp(-\alpha t) + B\exp(-\beta t), \qquad (14.73)$$

with a value of $R^2 > 0.99$ as depicted in Figure 14.13.

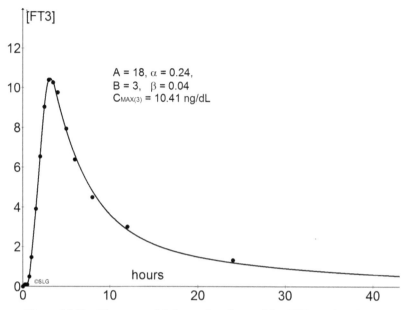

Figure 14.13 Bi-exponential decay after C_{max} of the [FT3] profile of P01.

14.5 Discussion

The investigation about the properties of appearance, distribution, and half-life decay of levothyroxine and liothyronine has demonstrated that the GIT has a pronounced influence on the appearance characteristic. After the ingestion of either L-T4 or L-T3, the appearance is accurately modeled with the single-compartment approach resulting in one dominant time constant. Although it seems unimportant how the drug appearance manifests itself, this analysis had the intention to validate the chosen model of a linear rising drug concentration as a function of time and the accuracy of the resulting appearance characteristic. The effect of the double integration has been demonstrated.

The drug decay section is important because we could model this behavior with a two-compartment approach [8]. The modeling, based on an electric circuit with resistors and capacitors, is of interest because we can verify the results with actual electrical circuit hardware as well as using *in silico* simulation with a standard electrical circuit simulator like Simetrix [4].

The relatively strong decay of the drug concentration in the distribution phase directly after the maximum C_{max} plays a major role in the analysis of the pulsatile concentration behavior of [TSH] and [FT4] in Chapter 17 about the circadian feedback dynamics. Modeling with electrical circuits provides a wide variety of possibilities in physiology.

The moment we find a validated result, we can have a rapid confirmation that the circuit model accurately models and describes the underlying physiological reality.

With this modeling approach we can also demonstrate the individual differences in the appearance time constant and the variability in T4 and T3 half-life behavior.

References

[1] Urso, R., Blardi, P., Giorgi, G. (2002). A short introduction to pharma-cokinetics. *Eur. Rev. Med. Pharmacol. Sci.* 6, 33–44.

[2] Goede, S. L., Leow, M. K., Smit, J. W. A., Dietrich, J. W. (2014). A novel minimal mathematical model of the hypothalamus–pituitary–thyroid axis validated for individualized clinical applications. *Math. Biosci.* 249, 1–7.

[3] Jonklaas, J., Burman, K. D., Wang, H., and Latham, K. R. (2015). Single-dose T3 administration: kinetics and effects on biochemical and physiological parameters. *Ther. Drug Monit.* 37, 110–118.

[4] SIMetrix Technologies Ltd (2015). *Circuit Simulation, 78 Chapel Street, Thatcham, Berkshire, RG18 4QN United Kingdom.* Available at: http:/www.simetrix.co.uk

[5] Hodgkin, A., Huxley, A., and Katz, B. (1952). Measurement of current – voltage relations in the membrane of Loligo. *J. Physiol.* 116, 424–448.

[6] Hodgkin, A. L., Huxley, A. F. (1952). A quantitative description of membrane current and is application to conduction and excitation in nerve. *J. Physiol.* 117, 500–544.

[7] Schenkman, A. L. (2005). *Transient Analysis of Electrical Power Circuits Handbook.* New York, NY: Springer. doi: 10.1007/0-387-28799-X

[8] Goede, S. L., Latham, K. R., Leow, M. K., and Jonklaas, J. (2017). High resolution free triiodothyronine thyrotropin (FT3–TSH) responses to a single dose of liothyronine in humans: evidence of distinct inter-individual differences unraveled using an electrical network model. *J. Biol. Syst.* 25, 119–143. doi: 10.1142/S0218339017500073

15

Pharmacokinetics of Liothyronine (L-T3) and Levothyroxine (L-T4)

"Declare the past, diagnose the present, foretell the future."

–Hippocrates (460 BC–370 BC)

15.1 Introduction

Today, millions of people are dependent on the use of daily oral levothyroxine (L-T4), and in rare cases supplemented with liothyronine (L-T3) due to permanent hypothyroidism of various etiologies associated with a range of incapacitated peripheral deiodinase functionality. A smaller fraction of hypothyroid people uses natural desiccated thyroid derived from porcine thyroids instead of synthetic L-T4 or L-T3. Like any drug and hormone, the plasma levels of ingested L-T4 are dependent on the biological processes of absorption, distribution, biotransformation, and clearance, i.e., pharmacokinetics (PK) [1, 2]. Given the enormous variety of medicines and myriad dosage schemes in this present age, it is a wonder why there is a dearth of PK analysis concerning thyroid hormone replacement with L-T4 and L-T3 in the available literature.

Thyroid hormone replacement by synthetic T4 and/or T3 aims to mimic normal thyroid hormone secretion in an ideal substitution therapeutic regimen. Because we have now developed a validated theory to determine the euthyroid HPT set point [3], we can apply the L-T4 substitution therapy to result in the targeted [FT4]–[TSH] values. We examine the PK model of the synthetic thyroid hormones levothyroxine, or thyroxine sodium (L-T4), and triiodothyronine or liothyronine sodium (L-T3) when administered orally. The presented theoretical framework in this chapter provides a clear insight into the PK effects of specific dosage schedules and deviations due to compliance issues, and also allows the derivation of clinically applicable correction measures necessary for rapid yet safe compensatory dosing to achieve the targeted euthyroid steady state. An important process of PK is the gradual accumulation of orally ingested L-T4 in the circulatory compartment which eventually leads to a steady-state level is dependent only on the amount ingested, bioavailability, dosing frequency, and half-life of the thyroid hormones [4].

L-T4 is available worldwide in several formulations. Depending on the proprietary L-T4 brands, these come in tablets of varying strengths, commonly ranging from 25 to 200 μg per tablet. An initial preliminary target dose is often calculated using one of several factors, such as body weight for instance [5]. Conventionally, a patient's subsequent thyroid hormone requirement can be confirmed via dose titration steps made over the course of several weeks during outpatient clinic visits based on thyroid function test responses (i.e., serum free thyroxine [FT4] and thyrotropin [TSH]) and the patient's state of well-being [6]. However, there is an increasing awareness of suboptimal outcomes using a "normal range" approach as described above to find a patient's euthyroid dose that corresponds to optimal health. We have published preliminary data that separately showed that a medication strategy based on a patient's individualized euthyroid homeostatic set point is more likely to lead to better clinical results [3].

Physiologically, besides T4, a healthy thyroid also produces continuously a steady amount of T3, which is about 20% of the total amount of T3 in the circulation [7, 8]. In L-T4 replacement therapy for patients whose thyroids have failed or been completely ablated, intra-thyroidal T3 is largely absent. This potential lack of T3 is compensated by $5'$-deiodinase acting on T4 to generate T3 peripherally in organs such as the liver [7]. However, there are rare cases of people with deficient $5'$-deiodinase activity in whom such a T3 compensatory mechanism will be insufficient [7, 9, 10]. These are cases where the use of additional L-T3 to their L-T4 regimen can be synergistic in achieving personalized euthyroidism. Besides these rare $5'$-deiodinase activity insufficiency, there is no rationale to prescribe additional T3 along with T4. This is because the production of T3 from a healthy thyroid provides only 20% of the total measurable [FT3] in plasma. The measured [FT3] in plasma is a residual reflection of the total deiodinase results, and therefore, the thyroid production of T3 is easily compensated by the overwhelming [FT4] to [FT3] conversion just by making the amount of [FT4] sufficiently available.

The half-life $(t_{1/2})$ of T4 is about 7 days in the euthyroid state which leads to relatively stable levels of accumulated T4. The half-life of T3 $(t_{1/2} = 21$ h) results in greater degrees of daily serum level fluctuations of T3 than is the case with T4. Clinical guidelines concerning the use of L-T4 and L-T3 in combination have important consequences when the indicated L-T3 dosages including a number of alternative dosage schemes are applied [7]. In the case of different dosage schedules for L-T3, we show the effects of a daily dose over 24 h of 12.5 μg L-T3. Besides this dosage schedule, we also analyze the situation with 12.5 μg L-T3 split into two doses of 6.25 μg L-T3 over 24 h and compare the results.

Finally, we propose a potentially applicable dosing design method of L-T4 intake for fast recovery to attain the desired steady state when compliance of L-T4 intake is imperfect, such as instances of doses missed consecutively for some days. The missed dose can generally be compensated in 1 or 2 days. A similar approach can be applied to obtain the desired steady-state level of FT4 within 1–2 weeks for medically suitable patients just started on L-T4 replacement therapy, instead of the standard period of 6–8 weeks. In the appendix, we show a number of dosing regimen design tables for different dosages that can be applied. The presented investigation can also serve as a reference for future guidelines for the general use of L-T4.

15.2 Thyroxine Kinetics

15.2.1 PK of L-T4 with T4 Level Dependent and Constant $t_{1/2}$ When Administered as a Daily Dose

The appearance process of L-T4 is based on incremental discrete steps of 1 day, 24 h. In this process, we assume that the effective uptake is 100 μg/day (i.e., bioavailability = 100%, or $F = 1.0$), after which the PK is monitored at intervals of 24 h. With the half-life of 7 days, we can expect the initially absorbed amount of 100 μg taken at the beginning of the day results in a remaining amount of 90% after 24 h. The next day a new dose of 100 μg is added and so on and so on until the daily administered amount equals the metabolized amount over that period. When the dose equilibrium has been reached, we define this as the steady-state level.

15.2.2 Derivation of the Steady-State Level

From Figure 15.1, we can easily appreciate that the accumulation of T4 behaves as an asymptotically accumulating characteristic saturating at a steady-state level from which no higher accumulation will occur. The steady-state level A_e can be derived as follows. At steady state, the metabolic loss over 24 h equals the daily dose. With the assumption of a daily intake of

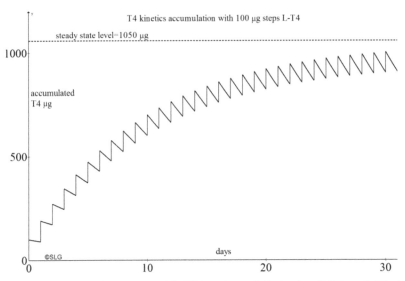

Figure 15.1　Accumulation process of L-T4 based on a daily intake of 100 µg L-T4. After about 40 days, equilibrium will be reached indicated as steady state level of accumulated L-T4.

100 µg L-T4, the decay over 24 h is

$$A_e\{1 - \exp(-0.1)\} = 100 \ \mu g \tag{15.1}$$

Then,

$$A_e = \frac{100 \ \mu g}{1 - \exp(-0.1)} = 1050 \ \mu g \tag{15.2}$$

The generalized form for the steady-state level for a medication is based on discrete periodical doses with dosing interval of n days.

$$A_e = \frac{\text{Periodical dose in } \mu g \text{ over } n \text{ days}}{1 - \exp(-n\delta)} \tag{15.3}$$

This is the amount directly after the uptake of the daily dosage and can be considered as the top value of the time-dependent function of the T4 level. For an equilibrium situation, we find the following signal form as depicted in Figure 15.2.

The continuous-time function of the steady-state value can be written as

$$A_e(t) = D_d + (A_e - D_d)\{1 - \exp(-n\delta)\}, \tag{15.4}$$

with D_d representing the daily dose and $\delta = 1/\tau$.

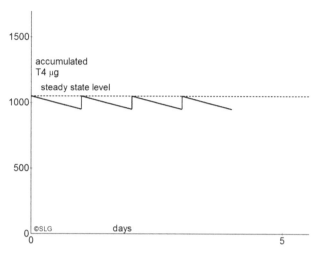

Figure 15.2 Equilibrium situation with top level A_e and the end of the period level $A_e \exp(-n\delta)$ where $\delta = 1/\tau$ or $\delta = 0.1$.

15.3 Average Value of Time-Dependent Functions

From Figure 15.2, we can derive the average T4 level:

$$T4_{average} = \frac{1}{T} \int_0^T A_e \exp(-\delta t)dt, \tag{15.5}$$

with dosing interval $T = 1$ (1 day)

$$T4_{average} = -\frac{A_e}{\delta} \left[\exp(-\delta t)\right]_0^1, \tag{15.6}$$

then

$$T4_{average} = \frac{A_e}{\delta} \left\{1 - \exp(-\delta)\right\}. \tag{15.7}$$

For

$$A_e = \frac{\text{Daily dose}}{1 - \exp(-\delta)} \tag{15.8}$$

we find

$$T4_{average} = \frac{\text{Daily dose}}{\delta} \tag{15.9}$$

Thus, for the daily dose of 100 µg, we find

$$T4_{average} = \frac{\text{Daily dose}}{\delta} = \frac{100}{0.1} = 1000 \, \mu g \tag{15.10}$$

15.4 RMS Value of Continuous Time-Dependent Functions

From the previous part, we described the average value of a time-dependent function.

The varying part of the signal is relatively small compared to the part under the baseline value. With the method of root-mean-square (RMS) value, we can determine the effect of a signal. As an example, we use the direct current voltage V_{DC} to light up an incandescent light bulb. When we use the AC power grid of 230 V with a sinusoidal signal form, the value of 230 V indicates directly the RMS value of the power grid. The peak value of the sinusoidal voltage is determined as: $230\sqrt{2} = 325$ V.

This means that a sinusoidal peak voltage of 325 V has the same effect with respect to the light intensity output by the incandescent bulb as a direct current voltage of 230 V.

We can derive that V_{RMS} of $A\sin(x)$ equals $A/\sqrt{2}$ [11].

We will use the following definition:

$$A_{RMS} = \sqrt{\frac{1}{T} \int_0^T [f(t)]^2 dt} \tag{15.11}$$

In this example, we have,

$$f(t) = A\sin(\omega t), \tag{15.12}$$

with

$$\omega = \frac{1}{T}, \tag{15.13}$$

where T represents the period time depicted in Figure 15.3.

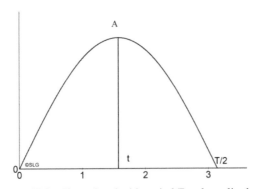

Figure 15.3   Sinus signal with period T and amplitude A.

It is sufficient that we investigate the RMS value over half a period $T/2$ and write $[f(t)]^2$ as

$[f(t)]^2 = \{A\sin(\omega t)\}^2 = A^2\sin^2(\omega t)$ for the interval $0 < t < T/2$,

then

$$\{A_{RMS}\}^2 = \left(\frac{2A^2}{T}\right)\int_0^{T/2}\sin^2(\omega t)dt, \qquad (15.14)$$

resulting in

$$\{A_{RMS}\}^2 = \frac{2A^2}{T}\left[\frac{1-\cos(2\omega t)}{2}\right]_0^{T/2} = \frac{A^2}{2} \qquad (15.15)$$

For the RMS value of A, we then find,

$$A_{RMS} = \frac{A}{\sqrt{2}} \approx 0.7A \qquad (15.16)$$

The average value of $A\sin(\omega t)$ over the interval $0 < t < T/2$ is defined as:

$$A_{AV} = \frac{2}{T}\int_0^{T/2}A\sin(\omega t)dt = \frac{2A}{T}[-\cos(\omega t)]_0^{T/2} = \frac{2A}{\pi} \approx 0.64A$$
$$(15.17)$$

From this example, we find that the RMS value is larger than the average value.

15.5 Weekly Administration of L-T4 at Seven Times the Normal Daily Dose

Under certain conditions such as issues relating to poor compliance, a patient may be administered a weekly dose of L-T4 instead of daily dosing. This dose is based on a seven times the daily dose of an individual. Such a strategy has been shown to be practical and safe [12]. However, it should be reiterated that this is tolerated mainly because thyroxine is a prohormone. A weekly dosing using seven times the daily dose of L-T3 would not work and might even be lethal as T3 is extremely potent.

When we use the reference example of a 100 µg daily L-T4, a weekly dose means ingesting 700 µg L-T4 at once on a specific day of the week. This dosage procedure is the weekly repeated. Calculating the steady state of T4 with Equation (15.2), we find a value of 1,390 µg after about 6 weeks. The PK effect of this approach is depicted in Figure 15.4. This is more than

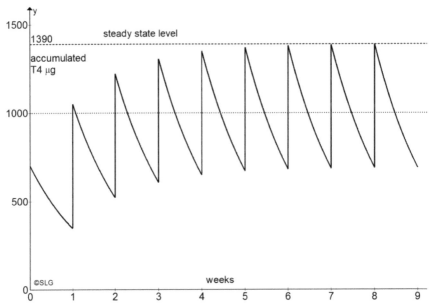

Figure 15.4 Appearance and clearance of T4 with a 700 μg weekly dosage resulting in a steady-state value of $A_e = 1,090$ μg and an average value $T4_{average} = 780$ μg.

the level of 1,050 μg which would be accomplished in about the same time using 100 μg daily. However, the average value drops to 780 μg. From this example, we see a relatively large time-varying effect in [T4] and belonging [FT4] of which the individual [TSH] values can be calculated as a function of time.

Applying the RMS definition of (Equation 15.11), we find the signal form of Figure 15.3.

In the following, L-T4 as a function of time is written as:

$$F_{L-T4}(t) = 1390 \exp(-0.1t) \tag{15.18}$$

$$F_{L-T4}(t)_{RMS} = \sqrt{\frac{1}{7} \int_0^7 (1390)^2 \exp(-0.2t)dt} = 799.6 \tag{15.19}$$

This result shows that the RMS value of $F_{L-T4}(t)$ is significantly higher than the average value and that the RMS value should be taken into account for these situations. *The effective value is the same as if the thyroid gland had produced T4 continuously over the period of 1 week.*

15.6 Triiodothyronine Kinetics

15.6.1 Accumulation of T3 Based on Various Dosages of Daily Liothyronine (L-T3)

Hypothyroidism is generally treated by means of L-T4 monotherapy. However, in some cases, the peripheral expression of $5'$-deiodinase may be insufficient to compensate for T3. This is thus a justifiable indication for the addition of triiodothyronine or liothyronine (L-T3) with levothyroxine (L-T4) [10]. We derive accumulation schemes based on dosages of L-T3 12.5 and 6.25 µg. Other dosage schemes (e.g., two times per 24 h) will be discussed regarding T3 accumulation effects over 24 h which is similar to that of T4 with the main difference of a much smaller $t_{1/2}$. T3 has an estimated $t_{1/2}$ of about 21 h which equals $21/24$ of a day. We calculate the value of the time constant of T3 from:

$$\tau = \frac{t_{1/2}}{\ln(2)} = 1.4427 t_{1/2} = 1.2 \text{ days,} \tag{15.20}$$

with

$$\frac{1}{\tau} = 0.792 \tag{15.21}$$

This results in the following expression for the decay over 24 h. For simplicity, we round 0.792 to the value of 0.8:

$$\delta = \frac{1}{\tau} = 0.8 \tag{15.22}$$

$$A_t = A_0 \exp(-0.8t) \tag{15.23}$$

15.6.2 Accumulation of T3 over a Period of 9 days with a Daily Dose D_d of 12.5 µg L-T3

Table 15.1 gives an overview of the accumulation process.

The steady-state level for this dose is:

$$A_e = \frac{D_d}{1 - \exp(-\delta)} = \frac{12.5}{1 - \exp(-0.8)} = 22.7 \text{ µg} \tag{15.24}$$

$$T3_{average} = \frac{D_d}{\delta} \tag{15.25}$$

Table 15.1   Accumulation of T3 over 9 days with a daily dose D_d of 12.5 μg L-T3

Day Number	Accumulated Amount of T3 (μg)	Remains T3 after 24 h (μg)
0	12.5	5.6
1	18.1	8.1
2	20.6	9.3
3	21.8	9.8
4	22.3	10.0
5	22.5	10.1
6	22.6	10.2
7	22.7	10.2
8	22.7	10.2

For comparison, the effect of a normal operating thyroid is based on a continuous stream of secreted T3 which is equivalent to a continuous amount of infinitesimally small increments resulting in an average value of the time-dependent T3 based on discrete daily increments [13]. The continuous accumulation function for the T3 steady-state level $T3_s$ is then:

$$A_e(t)_{T3} = D_d + \{A_e - D_d\}\{1 - \exp(-\delta t)\} \qquad (15.26)$$

From Figure 15.5(a), we appreciate that the daily variations in L-T3 concentrations are in the same order as the daily dose, here about 12.5 μg per 24 h. These relatively large fluctuations can be very uncomfortable and basically avoidable when the intake frequency of the dose is higher and the dosage is proportionally smaller. In the clinical situation that L-T3 is deemed advantageous to be co-administered with L-T4, we can either use the higher frequency strategy or else make use of a slow release L-T3 tablet [14] if such a drug formulation is available. However, we have to realize that the split dosage of 6.25 μg L-T3 given 12 hourly has a different effect in A_e from the case of a single dosage of 12.5 μg L-T3 per 24 h, but results in the same average level of T3. Because the amplitude of 24 h fluctuations of the 12.5 dosage scheme is rather pronounced, the 2012 ETA Guidelines introduce alternative schemes based on 6.25 μg L-T3 ingested twice per 24 h [10]. This situation is listed in Table 15.2 and will be analyzed as is depicted below.

The expected steady-state level is then

$$A_e = \frac{\text{Periodical intake in μg}}{1 - \exp(-\delta n)} = \frac{6.25}{1 - \exp(-0.4)} = 19.0 \; \mu g (n = 0.5)$$

$$(15.27)$$

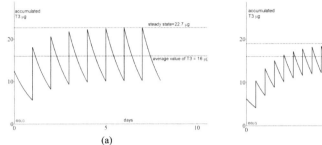
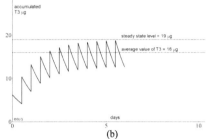

Figure 15.5   (a) 12.5 mg L-T3 per 24 h and (b) 6.25 mg L-T3 per 12 h.

Table 15.2   Liothyronine uptake based on 6.25 µg L-T3 taken twice within 24 h

Day Number	Accumulated Amount of T3 (µg)	Remains T3 after 24 h (µg)
0	6.25	4.2
0.5	10.4	7.0
1	13.2	8.9
1.5	15.1	10.1
2	16.4	11.0
2.5	17.2	11.6
3	17.8	11.9
3.5	18.2	12.2
4	18.4	12.4
4.5	18.6	12.5
5	18.7	12.6
5.5	18.8	12.6

Dosing result of 12.5 µg L-T3 per 24 h (Figure 15.5a) and dosing result of 6.25 µg L-T3 twice daily.

Figure 15.5(a) shows that the 12.5 µg dosage of L-T3 over 24 h results in a relatively large T3 ripple over the same period. This variation is reduced with the dosage scheme of 6.25 µg L-T3 administered twice daily over 24 h as depicted in Figure 15.5(b). It is expected that the value of A_e in the 12.5 µg L-T3 daily situation is higher than 6.25 µg L-T3 twice daily, but the average value in both situations is the same. We get similar results when we compare the accumulative effects of 6.25 µg L-T3 over 24 h with an alternative dosage scheme of 3.125 µg twice daily within 24 h. As before, the average level remains unchanged but the daily variations are reduced by a factor two using the alternative dosing scheme.

15.7 Compensation Strategies When L-T4 Is Not Taken Regularly

For a variety of reasons, one or more daily doses L-T4 are often omitted by patients. Sometimes, this could be a 1-day omission after which the compensation is very simple and generally safe, as can be shown to be achieved by taking twice the normal dosage the next day. When this corrective step is not taken and the daily routine resumes according to the prescribed L-T4 "a pill once-daily" regimen, it will actually take *several weeks* to reach the target normal steady-state level again, though the difference if anything is admittedly clinically insignificant for most instances due to the long $t_{1/2}$ of T4. The PK effect of the omission of 1 day of L-T4 is depicted in Figure 15.6.

Table 15.3 shows the numerical result of omitting a single daily dose after which the normal daily intake regime is resumed.

15.7.1 Acceleration Toward Steady State for a 100 μg L-T4 Daily Dosing Scheme Within 1 Week

The desired steady-state plasma level of 1,050 μg based on a normal daily schedule of 100 μg L-T4 can easily be reached in a shorter period when the initial dosage steps are allowed to be increased to a daily dose of 300 μg L-T4. Table 15.4 shows the results which are depicted in Figure 15.7.

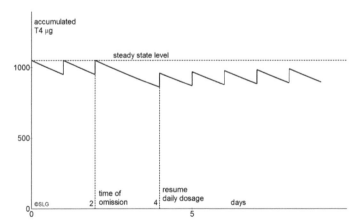

Figure 15.6 The situation after omission of a single dose of daily 100 μg L-T4 and then resuming with the usual normal dose without any compensatory dosing. The diagram shows the effect of slow increase to the steady-state level.

Table 15.3 Intake omission for 1 day after which the accumulation with 100 μg/day is resumed

Day Number	Accumulated Amount of T4 (μg)	Remains T4 after 24 h (μg)
0	1050	950.2
1	950.2	860.0
2	960.0	868.6
3	968.6	876.5
4	976.5	883.6
5	983.6	890.0
6	990.0	896.0

Table 15.4 Rapid escalation to the desired steady-state level of 1,050 μg T4 in 4 days

Day Number	Accumulated Amount of T4	Remains T4 after 24 h
0	300	271.5
1	571.5	517.1
2	817.1	739.7
3	1039.7	940.7
4	1050.7	950.1
5	1050.1	950.1

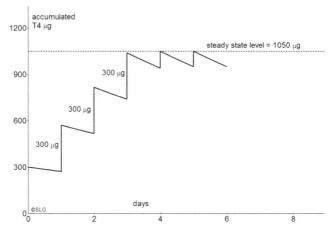

Figure 15.7 Illustration of the accelerated accumulation of T4 toward steady state using thrice the daily dose of an originally 100 μg L-T4 dosage schedule to compensate for missed L-T4 doses.

Starting the first dose on day 0, the second dose on day 1, etc., we find after four daily doses of 300 μg L-T4 that the steady-state level is practically reached. At the beginning of day 4, we can administer 110 μg L-T4, and the following day, we can continue with 100 μg L-T4 on a daily basis.

Figure 15.7 depicts the rapid accumulation to the desired steady-state level with four daily doses of 300 µg L-T4, starting at day "0" followed by the normal daily dosage of 100 µg L-T4. This is purely for illustrative purposes. We stress here that the maximum allowable dosage obviously depends on any contraindication to supranormal doses of L-T4, prevailing guidelines, recommendation from regulatory authorities on dosing (e.g., FDA, EMA, HSA, etc.), risk–benefit ratio, and the health status of the patient which must be monitored appropriately if such a regimen is ever adopted. The usefulness of such an accelerated dosing method is that it can significantly shorten the period to reach the desired steady state [FT4] after some daily doses have been missed. This can translate into faster resolution of hypothyroid symptoms in certain clinical scenarios.

Based on the mathematical pharmacokinetic model as illustrated above, we can design the corresponding L-T4 dosing compensation regimen when one or more daily doses of L-T4 supplement have been missed. The effect of one missed dose is indicated in Figure 15.8. Alternatively, we can utilize the same technique of dosing when there is a clinical indication to elevate the steady-state level to a higher value. The number of days it takes to reach this level is dependent on the maximum tolerable daily dose without incurring toxicity in any specific individual. We discuss these strategies with practical examples as shown below. In the following, a general method for L-T4 dosing compensation will be discussed.

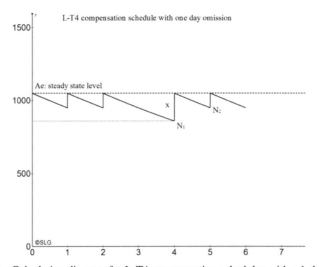

Figure 15.8   Calculation diagram for L-T4 compensation schedules with a 1-day omission.

Case 1: In the case that a dosage has been omitted, for one or more days, the value of T4 will drop to a certain level N_1. To restore the original value of the steady-state level, the difference between N_1 and A_e equals Δ and has to be bridged.

The general expression for the number of days φ that L-T4 is omitted and results in

$$n = 1 + \varphi \tag{15.28}$$

According to the decay process, we find

$$N_1 = A_e \exp(-\delta n) \tag{15.29}$$

$$\Delta = A_e - N_1 = A_e - A_e \exp(\delta n) = A_e\{1 - \exp(\delta n)\} \tag{15.30}$$

The daily decay is represented by

$$A_e\{1 - \exp(-\delta)\} \; n = 1 \;\; \text{and} \;\; \delta = 0.1, \tag{15.31}$$

with $\delta = \frac{1}{\tau} = 0.1$

Furthermore, we define the factor ρ as $\rho = \exp(-\delta)$ where $\rho = 0.9$ for L-T4.

In Figure 5.2, the difference between the steady-state level A_e and the last attained level $N1$ equals Δ and has to be compensated. This can be accomplished by taking the dosage Δ directly. The value of Δ as expressed with (Equation 15.30) results in:

$$\Delta = A_e - N_1 = A_e\{1 - \exp(\delta n)\} = 190.3 \; \mu g \tag{15.32}$$

Here, the daily dosage is omitted for 1 day resulting in a value for n: $n = 2$

Case 2: If the amount of Δ is too high to compensate at once due to toxicity limitations for the specific individual, the level A_e can be reached in a 2-day scenario as follows.

We can appreciate the analysis situation from Figure 15.9.

We will divide the gap Δ over a period of 2 days ($n = 3$) where x represents the bridging dosage. Thus, on day 5, we take the amount of x and obtain the new level $N_1 + x$. The decay amount for L-T4 over 24 h is then:

$$N_2 = \rho(N_1 + x) \tag{15.33}$$

The new difference between N_2 and A_e results in:

$$A_e - N_2 = x, \tag{15.34}$$

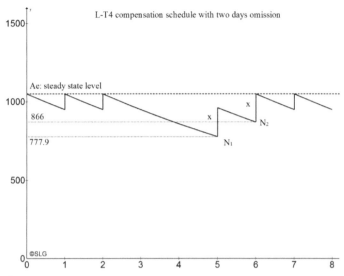

Figure 15.9   Compensation schedule of L-T4 for the omission of two consecutive daily doses.

and then it follows that

$$A_e - \rho N_1 - \rho x = x, \qquad (15.35)$$

or

$$x + \rho x = A_e - \rho N_1, \qquad (15.36)$$

from which we can derive

$$x = \frac{A_e - \rho N_1}{1 + \rho} \qquad (15.37)$$

Over a period of 2 days, x is the amount which has to be taken so as to attain the desired level A_e.

Case 3: If the amount of Δ is too high to be compensated at once, the level A_e can be reached in a 3-day scenario as follows. From Case 2, we have derived the value of N_2.

The situation for Case 3 is depicted in Figure 15.10. From day 2 to day 8, the intake of L-T4 is omitted. At day 8, the level N_1 has been reached. When the compensation dosage has to be divided over a period of 3 days, we have

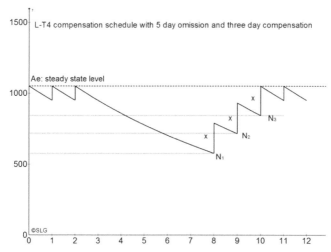

Figure 15.10 Calculation diagram for a 3-day L-T4 omission compensation dosage division schedule after an omission of 5 days. $\varphi = 5$ and thus $n = 6$.

the following situation. After reaching level N_2, at day 9, the amount x, as was taken the previous day, is added to the pool. This results in:

$$N_2 = \rho(N_1 + x) \tag{15.38}$$

After 24 h, this amount will be reduced to N_3 on day 10.

$$N_3 = \rho(N_2 + x) = \rho\{\rho N_1 + \rho x + x\} = \rho^2 N_1 + \rho^2 x + \rho x \tag{15.39}$$

The remaining difference of A_e and N_3 results in

$$A_e - N_3 = x \tag{15.40}$$

from which we can calculate x again as:

$$A_e - N_3 = A_e - \rho^2 N_1 - \rho^2 x - \rho x = x \tag{15.41}$$

$$x(1 + \rho + \rho^2) = A_e - \rho^2 N_1, \tag{15.42}$$

resulting in

$$x = \frac{A_e - \rho^2 N_1}{(1 + \rho + \rho^2)} \tag{15.43}$$

We then can write a generalized dosage equation divided over "m" days as

$$x = \frac{A_e - (\rho^{(m-1)})N_1}{\sum\limits_{k=1}^{k=m} \rho^{(k-1)}} \tag{15.44}$$

With the expansion of A_e and N_1, we find

$$x = \frac{\left(\frac{\text{Daily dose}}{1-\exp(-\delta)}\right)\{1 - (\rho^{(m-1)})\exp(-\delta n)\}}{\sum\limits_{k=1}^{k=m} \rho^{(k-1)}}, \tag{15.45}$$

with "m" representing the number of days that should be used in the division of the gap to compensate for the loss of T4. After the steady-state level has been reached, the daily dosage for which the steady state was originally targeted and designed has to be resumed as per usual routine.

15.8 Calculation of a Reduced Dosing Scheme

The previous sections analyzed the fast route to a starting or enhanced steady state of L-T4 or L-T3. The following will show methods for calculation to a reduced steady-state level. We define the current steady state as A_1 and the reduced steady state as A_2.

The natural decay is defined as:

$$A(t) = A_1 \exp(-\delta t) \tag{15.46}$$

The time necessary to reach $A(t) = A_2$ is expressed as

$$A_2 = A_1 \exp(-\delta t), \tag{15.47}$$

from which follows

$$\exp(-\delta t) = \frac{A_2}{A_1}, \tag{15.48}$$

resulting in

$$-\delta t = \ln\left(\frac{A_2}{A_1}\right), \tag{15.49}$$

or

$$t = -\frac{1}{\delta}\ln\left(\frac{A_2}{A_1}\right) \tag{15.50}$$

Because "t" is applied as an integer, we round the value of "t" to the nearest higher value.

This results in the lower value A_2'. The difference $\Delta_A = A_2 - A_2'$ will be the first dose to reach the targeted steady-state level A_2. The following day will be resumed with:

$$\text{Daily dose} = A_2\{1 - \exp(-\delta)\} \tag{15.51}$$

15.9 Negative Long-term Effects of Externally Administered T3

The L-T4 replacement therapy is actually a very good approximation of the normal [FT4] variations encountered in healthy persons with an operational thyroid. The thyroidal secretion of T3 in a normal situation contributes for only 20% of the [FT3] encountered in plasma concentrations. Eighty percent of the [FT3] in plasma is the result of peripheral 5′-deiodination. From a biological systems point of view, this thyroidal T3 contribution is not necessary compared to the deiodinase capacity of the peripheral tissues. We can argue that the use of externally administered T3 in the form of L-T3 continuously for an indefinite period of time is not recommended.

The negative long-term effects of externally administered T3 include the following.

1. Externally administered T3 leads to a back regulation of peripheral deiodinase processes [15].
2. Chronic T3 overdosage can be toxic at the tissue level.
3. With local hypothalamic [FT3] predominating over [FT4] as the feedback signal to the hypothalamus–pituitary system, the [TSH] inhibition is now controlled by [FT3]. This contributes to difficulty in achieving a stable point of equilibrium around a stable value of [TSH] [13].
4. As a corollary from point 3, the [FT4] set point is shifted to a sub-level. Figure 15.11 demonstrates the effect of competing concentrations of [FT3] in the hypothalamus–pituitary characteristic when L-T3 is used in a combination replacement therapy with L-T4.
5. The application of externally administered T3 leads to violent variations in [FT3] plasma levels which cannot be controlled other than a regulated and controlled continuous flow of T3 fully according to the behavior of a normal thyroid. Any other approach leads to an hour-by-hour dosage system to keep the [FT3] concentration variations to an acceptable level.

Because of the short time of several hours that the taken L-T3 dose is peaking well beyond the average plasma value of [FT3], generally more than 1.6 times the plasma value, we will find the effect of a shifted HP characteristic to the lower end of the [FT4] scale after one week of such a regimen.

The effect of the shift indicated in Figure 15.11 can be described by the application of the extended model:

$$[TSH] = \frac{M}{\exp(\varepsilon[FT3])\exp(\varphi[FT4])} \tag{15.52}$$

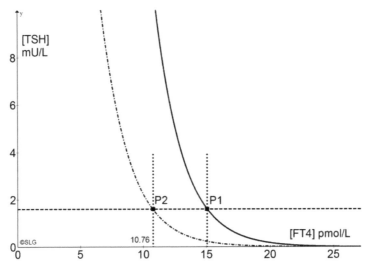

Figure 15.11   Shift of HP characteristic from mono L-T4 replacement therapy with set point P1 to the shift of the set point position P2 as a result of L-T4–L-T3 combination therapy.

From the expression of (Equation 15.52) and the value of ε, the competitive amount of [FT4] can be calculated.

The normally rather stable value of HP model parameter S can be reformulated as

$$S = M/\exp(\varepsilon[FT3]) \qquad (15.53)$$

In this case the inhibitory effect on hypothalamic TRH is enhanced by the externally administrated L-T3 resulting in a reduced value of S.

The new model parameter M is then directly associated with the amount of [TRH].

15.10 Discussion

The daily dosing of L-T4 should be complied strictly and regularly because missed doses often result in plasma thyroid hormone variations that in turn undermine euthyroidism and the state of health. Currently, the most important medication for thyroid hormone replacement is levothyroxine (L-T4) with a half-life of T4 of about 7 days resulting in daily variations of 10% of the steady-state level. The half-life of T3 is about 21 h which results in daily T3 fluctuations of about 50% of the steady-state level which is in the same order as the daily dosage. The daily T3 variations can be reduced by an alternative divided dosage scheme per 24 h. This results in the same average value of T3 but the variation is reduced to about 25% of the steady-state level.

In the previous analysis, the dependence of the actual T4 level on the half-life of T4 has been modeled and investigated. Besides the advantage of the extended half-life value during build-up in the low ranges of T4, the extended half-life protects the organism from a sharp decline of T4, which could otherwise possibly result in a deep hypothyroid state. Furthermore, we demonstrate that the average value of the time-varying concentrations of T4 and T3 represents the actual hormone levels that are currently accomplished by replacement therapy.

The above-discussed theoretical framework provides physicians to apply these results to advise their patients on the safe and recommended compensatory regimens to resolve plasma T4 levels due to imperfect compliance so as to attain the desired steady-state level in the minimum time.

Similarly, the same dosage strategy can be used to reach the desired steady-state level from the beginning of an L-T4 replacement therapy in 1 or 2 weeks instead of the normally practiced titration period of 6 weeks. Sometimes a reduced dosage regimen is desired. This is accomplished in a controlled way as described in Section 15.8, with a clear path to the new, lowered steady-state level. Although it might seem that the estimation of a compensation dosage is straightforward and intuitive, this is true only for the omission situation of 1 or 2 days. A closer analysis of this compensation mechanism is more complicated and relies on a deeper underlying theoretical framework. The objective is to restore the previous steady state in the shortest possible time without toxicity effects of L-T4.

In the appendix, a number of practical dosage schemes and the belonging compensation dosages are available in a tabular format. The compensation dosage is indicated as "x" over a bridging period of "m" days in the extreme right positioned column. When the bridging gap after two or more days of L-T4 omission is higher than the dose limit that the physician or specialist deems to be safe for the patient, dividing the dose over two or more days becomes an option for which the computation may be aided by the algorithm given in (15.45).

It has also been demonstrated that the use of L-T3 in the current tablet form induces a violent [FT3] surge during 2 to 3 hours well above 1.6 times the average [FT3] plasma level [16]. This effect has clearly been described when compared to the effect of [FT3] infusions with normal FT3 concentrations versus supra normal (1.6 × normal).

Only when this infusion with enhanced concentration, which compares well with the physiologic action of the thyroid, was given to hypothyroid subjects, after one week a euthyroid condition of the hypothalamic [FT3] could be established.

This explains the reduction of the [TRH] to the pituitary when L-T3 is taken as tablets and results then in the described shift of the HP characteristic. This effect can be confusing for thyroidologists or physicians because the [FT4] will then be appreciated at a significantly lower value with the same [TSH]. Many practitioners may then erroneously conclude that the patient needs more L-T4.

When the correct set point value of [TSH] is known, this can be used as the [TSH] target value when a patient is treated in combination with L-T3.

Appendix

L-T4 Compensation for Practical Situations

This appendix contains six dosing examples with the corresponding compensation algorithms. The indicated dosages are representative for most patients. The amounts for A_e are rounded for convenience. The division number that is the number of days (m) chosen to disperse the compensation amount is based on the individual case of maximum allowable dosage per 24 h.

$$\text{Dosage scheme } A_e = \frac{\text{Daily dosage}}{1 - \exp(-\delta)}\,\mu g,$$

with $n = 1 + \varphi$, $\rho = 0.9$, $\Delta = A_e - N_1$, and "m" the number of days over which the compensation amount is divided.

$$\text{Compensation dose } x = \frac{A_e - (0.9^{(m-1)})N_1}{\sum\limits_{k=1}^{k=m} 0.9^{(k-1)}}$$

Appendix 15.1 Omission compensation on daily dosage 75 μg

Day (n)	$\exp(-\delta n)$	Omitted Amount of Days (φ)	Remaining Amount $(N_1\,\mu g)$	Total Compensation Amount Δ in μg	Dividing Advise over "m" Days	Compensation Dosage Divided Amount x in μg
1	0.905	0	713	75	One day	75
2	0.819	1	645	143	One day	145
3	0.741	2	584	204	Two days	138
4	0.670	3	528	260	Two days	164
5	0.607	4	478	310	Two days	188
6	0.549	5	433	355	Three days	161
7	0.497	6	392	396	Three days	172
8	0.450	7	355	433	Three days	184

$A_e = 788\ \mu g$ remaining amount $N_1 = A_e \exp(-\delta n)$ $\Delta = A_e - N_1$

Appendix 15.2   Omission compensation on daily dosage 87.5 µg

Day (n)	exp $(-\delta n)$	Omitted Amount of Days (φ)	Remaining Amount $(N_1$ µg)	Total Compensation Amount Δ in µg	Dividing Advise over "m" Days	Compensation Dosage Divided Amount x in µg
1	0.905	0	832	88	One day	88
2	0.819	1	753	166	One day	166
3	0.741	2	682	238	Two days	161
4	0.670	3	616	303	Two days	192
5	0.607	4	558	362	Two days	244
6	0.549	5	505	415	Three days	189
7	0.497	6	457	463	Three days	204
8	0.450	7	414	506	Three days	216

Daily dosage 87.5 µg $A_e = 920$ µg remaining amount $N_1 = A_e \exp(-\delta n)$ $\Delta = A_e - N_1$

Appendix 15.3   Omission compensation on daily dosage 100 µg L-T4

Day (n)	exp $(-\delta n)$	Omitted Amount of Days (φ)	Remaining Amount $(N_1$ µg)	Total Compensation Amount Δ in µg	Dividing Advise over "m" Days	Compensation Dosage Divided Amount x in µg
1	0.905	0	950	100	One day	100
2	0.819	1	860	190	One day	190
3	0.741	2	777	273	Two days	185
4	0.670	3	704	346	Two days	219
5	0.607	4	637	413	Two days	251
6	0.549	5	577	473	Three days	215
7	0.497	6	522	528	Three days	232
8	0.450	7	473	577	Three days	247

Daily dosage L-T4 100 µg $A_e = 1050$ µg remaining amount $N_1 = A_e \exp(-\delta n)$

Appendix 15.4   Omission compensation on daily dosage 125 µg L-T4

Day (n)	exp $(-\delta n)$	Omitted Amount of Days (φ)	Remaining Amount $(N_1$ µg)	Total Compensation Amount Δ in µg	Dividing Advise over "m" days	Compensation Dosage Divided Amount x in µg
1	0.905	0	1,188	125	One day	125
2	0.819	1	1,075	238	One day	238
3	0.741	2	973	340	Two days	230
4	0.670	3	875	438	Two days	277
5	0.607	4	797	516	Three days	246
6	0.549	5	721	592	Three days	215
7	0.497	6	653	660	Four days	196
8	0.450	7	591	722	Four days	262

$A_e = 1313$ µg remaining amount $N_1 = A_e \exp(-\delta n)$

Appendix 15.5   Omission compensation on daily dosage 150 μg L-T4

Day (n)	exp $(-\delta n)$	Omitted Amount of Days (φ)	Remaining Amount $(N_1 \mu g)$	Total Compensation Amount Δ in μg	Dividing Advise over "*m*" Days	Compensation Dosage Divided Amount x in μg
1	0.905	0	1,433	150	One day	150
2	0.819	1	1,291	285	One day	285
3	0.741	2	1,161	415	Two days	280
4	0.670	3	1,056	520	Three days	266
5	0.607	4	957	619	Three days	296
6	0.549	5	865	711	Four days	271
7	0.497	6	783	793	Four days	288
8	0.450	7	709	867	Five days	267

$A_e = 1576$ μg remaining amount $N_1 = A_e \exp(-\delta n)$

Appendix 15.6   Omission compensation on daily dosage 175 μg L-T4

Day (n)	exp $(-\delta n)$	Omitted Amount of Days (φ)	Remaining Amount $(N_1 \mu g)$	Total Compensation Amount Δ in μg	Dividing Advise over "*m*" Days	Compensation Dosage Divided Amount x in μg
1	0.905	0	1,672	175	One day	175
2	0.819	1	1,506	333	One day	333
3	0.741	2	1,363	476	Three days	271
4	0.670	3	1,232	607	Three days	310
5	0.607	4	1,116	723	Four days	293
6	0.549	5	1,010	829	Four days	316
7	0.497	6	914	925	Five days	298
8	0.450	7	828	1,012	Five days	312

$A_e = 1839$ μg remaining amount $N_1 = A_e \exp(-\delta n)$

References

[1] Bauer, L. A. (2014). *Applied Clinical Pharmacokinetics*, Third Edn. New York, NY: McGraw-Hill.

[2] Winter, M. E. (2010). *Basic Clinical Pharmacokinetics*, 5th Ed. Philadelphia, PA: Wolters Kluwer/Lippincott

[3] Leow, M. K., and Goede, S. L. (2014). The homeostatic set point of the hypothalamus-pituitary-thyroid axis – maximum curvature theory for personalized euthyroid targets. *Theor. Biol. Med. Model.* 11:3. doi: 10.1186/1742-4682-11-3

[4] Eisenberg, M., and Distefano, J. J. (2009). *TSH-Based Protocol, Tablet Instavbility, and Absorbtion Effects on L-T4 Bioequivalence, Thyroid, 2009,* Vol. 19, New Rochelle, NY: Mary Ann Liebert inc. doi: 10.1089/thy.2008.0148

[5] Sukumar, R., Agarwal, A., Gupta, S., Mishra, A., Agarwal, G., Verma, A. K., and Mishra, S. K. (2010). Prediction of LT4 replacement dose to achieve euthyroidism in subjects undergoing total thyroidectomy for benign thyroid disorders. *World J. Surg.* 34, 527–531.

[6] Goede, S. L., Leow, M. K., Smit, J. W. A., and Dietrich, J. W. (2014). A novel minimal mathematical model of the hypothalamus–pituitary–thyroid axis validated for individualized clinical applications. *Math. Biosci.* 249, 1–7. doi: 10.1016/j.mbs.2014.01.001

[7] Wiersinga, W. M., Duntas, L., Vadeyev, V., Nygaard, B., and Vanderpump, M. P. J. (2012). ETA Guidelines: The use of L-T4 + L-T3 in the treatment of hypothyroidism. *Eur. Thyroid J.* 1:55–71. doi: 10.1159/000339444.

[8] Henneman, G., Docter, R., Visser, T. J., Postema, P. T., and Krenning, E. P. (2004). *Thyroxine Plus Low-Dose, Slow-Release Triiodothronine Replacement in Hypothyroidism: Proof of Principle*, Vol. 14, New Rochelle, NY: Mary Ann Liebert, Inc.

[9] Duntas, L. H. (2005). Reassessment of combined LT4 and LT3 treatment for hypothyroidism: the prospects for slow-release T3 preparations. *Hormones* 4, 108–110.

[10] Sugawara, M., Lau, R., Wasser, H. L., Nelson, A. M., Kuma, K., and Hershman, J. M. (1984). Thyroid T4 $5'$-deiodinase activity in normal and abnormal human thyroid glands. *Metabolism* 1984, 332–336.

[11] Stein, J. Y. (2000). *Digital Signal Processing*. Hoboken, NJ: John Wiley & Sons, 17.

[12] Walker, J. N., Shillo, P., Ibbotson, V., Vincent, A., Karavitaki, N., Weetman, A. P., et al. (2013), A thyroxine absorption test followed by weekly thyroxine administration: a method to assess non-adherence to treatment. *Eur. J. Endocrinol.* 168, 913–917. doi: 10.1530/EJE-12-1035

[13] Celi, F. S., Zemskova, M., Linderman, J. D., Babar, N. I., Skarulis, M. C., Csako, G., et al. (2010). The pharmacodynamic equivalence of levothyroxine and liothyronine. A randomized, double blind, crossover study in thyroidectomized patients. *Clin. Endocrinol.* 72, 709–715. doi:10.1111/j.1365-2265.2009.03700.x.

[14] Pabla, D., Akhlaghi, F., and Zia, H. (2009). A comparative pH-dissolution profile study of selected commercial levothyroxine products using inductively coupled plasma mass spectrometry. *Eur. J. Pharma. Biopharm.* 72, 105–111.

[15] Werneck de Castro, J. P., Fonseca, T. L., Ueta, C. B., McAninch, E. A., Gabor, S. A., and Wittmann, M. (2015). Differences in hypothalamic type 2 deiodinase ubiquitination explain localized sensitivity to thyroxine. *J. Clin. Invest.* 125, 769–781.

[16] Lechan, R. M., and Fekete, C. (2005). Role of Thyroid Hormone Deiodination in the Hypothalamus. *Thyroid*, 15, 883–897.

16

Circadian Feedback Dynamics and Set Point Stability Analysis of the Hypothalamus Pituitary Thyroid System

"The rhythm of biological relations makes the absolute appear in the relativity of time and space and represents the swing of life."

– Nomis Edeog (1948–Present)

16.1 Introduction

A healthy thyroid secretes both $3,5,3',5'$-tetraiodothyronine or thyroxine (T4) and $3,5,3'$-triiodothyronine (T3) at a more or less continuous rate over 24 h. Compared to a half-life of 0.75 days for T3, the prolonged half-life of T4 of about 7 days is largely due to its nearly complete binding to plasma thyroid hormone-binding proteins, chiefly thyroxine-binding globulin (TBG), transthyretin (TTR), and albumin [1]. These transport proteins which shield the hydrophobic molecular moieties of T3/T4 from their aqueous environment buffer a stable circulating free T4 and free T3 pools which are available for cellular uptake. Free T4 concentration, [FT4], and free T3 concentration, [FT3], are both subject to homeostatic control by the HPT axis.

It is estimated that less than 20% of circulating T3 is produced by the thyroid gland, while the majority of circulating T3 is generated by peripheral conversion of T4 [2]. Normally, [FT4] is kept to a constant basal level by means of a negative feedback mechanism as a result of homeostatic control [3]. This is accomplished by thyroid stimulating hormone [TSH] produced in the pituitary. A negative feedback mechanism over the loop containing the thyroid and the HP unit ensures a homeostatic equilibrium between [FT4] and [TSH]. A rising level of [FT4] results in a decreased level of [TSH] and vice versa. The values of these concentrations, when the HPT system is in

233

homeostatic equilibrium, are defined as the homeostatic set point values of [FT4] and [TSH] [4]. The pair of homeostatic [FT4]–[TSH] values in healthy persons is defined as the euthyroid homeostatic set point value.

Besides the general maintenance of the homeostatic level of [FT4], the feedback loop has been reported as operational on short-term hourly regulation, resulting in pulsatile behavior of the [TSH] secretion [5–7]. In the subsequent sections, [FT3] is not considered to contribute directly from plasma concentrations to the feedback mechanism to maintain the stable level of [FT4]. T4 itself regulates its own production via 5′-deiodination to T3 both centrally and peripherally, and hence the entire physiologically observed negative feedback on T4 has already taken into account via the net intracellular T3 action at the HP unit [8]. Circulating [FT3] adds partly to the overall negative feedback regulation of both [FT3] and [FT4] as exemplified by cases of pure T3 toxicosis encountered in autonomously functioning thyroid adenomas. In the normal physiological situation, intracellular FT3 derived from locally deiodinated FT4 explains most of the observed inhibitory effects of the production and release of thyrotropin releasing hormone (TRH) in the hypothalamus and TSH in the pituitary. Notably, the inhibitory effect on the TRH neurons in the hypothalamus conditionally influences the downstream stimulation of the production of TSH in the pituitary and determines the set point value of [FT4] [3, 9].

The circadian variations of [TSH], [FT4], and [FT3] signals will be discussed in this chapter. Our modeling approach is then validated based on existing data available in the public domain from published literature [10]. We analyze the mentioned thyroid system control effects in three different layers of operation in healthy subjects that were sampled during 24 h and had TSH concentrations determined with a 10-min interval and FT4 and FT3 plasma concentrations with a test interval of 1 h. The data of all participants were analyzed individually to avoid convolution from generalized averages and other statistical processing.

These are the key conceptual definitions alluded to in the foregoing dissertation:

1. Homeostatic levels of [FT4]–[TSH] as consequential to the HPT feedback operation
2. Circadian regulation as the 24-h feedback control
3. Short-term feedback effects observable as continuous stable, constrained amplitude oscillations of [FT4] and [TSH] with a period of about 1 h [11].

This analysis leads us to these predictions and outcomes:

1. Subdivisions into three distinguishable categories based on [TSH] levels
2. Position of the homeostatic set point during 24 h
3. Performance qualifier for the thyroid operation.

Furthermore, a postulate about the origin and rationale of circadian variations will be discussed.

16.2 Oscillators

An oscillator can be regarded as a signal generator or signal source generating an alternating level of some variable as a function of time. Such a generator can be realized as an electronic piece of hardware, known in telecommunication systems, but other signal generation examples can be found in nature and specifically in our own physiology manifested by bio-rhythms. Because of the complex nature of our bio-rhythmic generators, the signal output is generally not a simple sinusoidal function of time but has the features analogous to a pulsatile waveform. Especially in the manifestation of circadian rhythms, we encounter generated signals in the form of triangular-shaped periodic variations as we encounter in the charge and discharge characteristics of a capacitor [12, 13].

We begin with the linear non-saturating harmonic oscillator as depicted in Figure 16.1.

The transfer function derived from Figure 16.1 can be written as:

$$\frac{C}{R} = \frac{A}{1 + A\beta} \tag{16.1}$$

When the phase contribution of β as a function of the frequency meets the condition of 180°, the output is in phase with the input R and the system

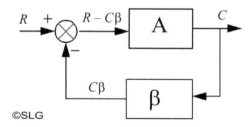

©SLG

Figure 16.1   Unity feedback configuration.

will oscillate at the frequency for which this phase shift will occur as long as the loop gain is larger than unity at that frequency. This type of oscillatory behavior can be encountered in the central nervous system or as metabolic dysfunctions well known and observed by physicians as tremors' characteristic of disorders as Parkinson's disease, benign essential tremors, hepatic encephalopathy, and even thyrotoxicosis.

Another common oscillator is known as relaxation oscillator. This type of oscillator is fundamentally based on non-linear signal behavior where the switch of one state to the other is signal level-dependent [12–14]. A realization example is given in Figure 16.2.

From Figure 16.2, we appreciate the central amplification unit A as the operational amplifier of the oscillator circuit. The time determining components are R_3, C, V_p, and V_n. The operational amplifier acts here as a level-dependent switch. When the voltage of the capacitor C, $U_c > U_a$, the output voltage U_0 equals V_n. The voltage U_{a1} then equals

$$U_{a1} = \frac{R_1}{R_1 + R_2} V_n, \tag{16.2}$$

and the capacitor will be discharged via R_3 to the most negative voltage V_n. This process continues until $U_c < U_a$ after which the operational amplifier switches to $U_0 = V_p$. Then we find for U_{a2}

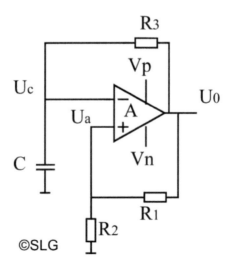

Figure 16.2   Relaxation oscillator with the operational amplifier A.

$$U_{a2} = \frac{R_2}{R_1 + R_2} V_p \qquad (16.3)$$

The total voltage difference over the capacitor is then U_d, written as:

$$U_d = U_{a1} - U_{a2} = \frac{R_1}{R_1 + R_2} V_n - \frac{R_2}{R_1 + R_2} V_p = \frac{R_1 V_n - R_2 V_p}{R_1 + R_2} \qquad (16.4)$$

The period of the relaxation signal is expressed as

$$T = R_3 C \left[\ln \left(\frac{2V_n - V_p}{V_n} \right) + \ln \left(\frac{2V_p - V_n}{V_p} \right) \right] \qquad (16.5)$$

In Figure 16.3, we simultaneously depict the switching signal U_0 and the capacitance voltage.

The signal forms depicted in Figure 16.3 are typical for many biological signals and apply specifically to the forms of circadian rhythms as discussed by Beersma et al. [12]. A variant of this form of oscillation will be modeled and discussed in Section 16.4.

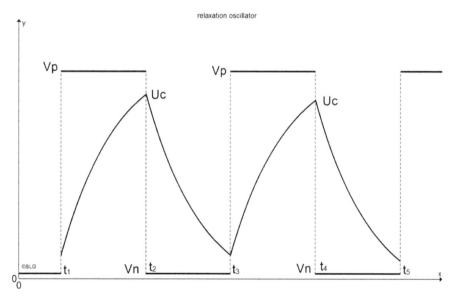

Figure 16.3 Signals of relaxation oscillator circuit of Figure 16.2. The signal time period $T = t_3 - t_1$.

16.3 Influence of [FT4] and [TSH] on Feedback Dynamics

The circadian variations of [TSH], [FT4], and [FT3] have been described and analyzed in various publications [7, 8]. In normal closed-loop situations, i.e., healthy persons, a circadian [TSH] rhythm is observable [15]. This is also observable in hypothyroid individuals who are substituted with levothyroxine (L-T4). In order to explain the circadian variations in the HPT hormone concentrations, we introduce the circadian clock as a base signal on which among others, the variations of [TSH] are superimposed as a result of the triangular charge–discharge signal [12, 13]. This mechanism is interpreted as the biological function of a T4 and T3 daily dosing process. The base signal of the circadian clock is presented in Figure 16.4 as drawn black line.

In general, we observe that the [TSH] variations are superimposed on the black, triangle-like, signal which will be presented later in various examples. The [TSH] signal is the result of sinusoidal [TSH] smaller local oscillations [5, 11] superimposed on the triangular signal to encourage the thyroid to

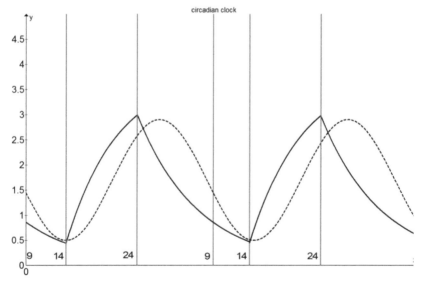

Figure 16.4 Circadian clock signal F_M (solid line) as the result of a pharmacokinetic charge and discharge process. The charging period is inferred from the shape and periodicity of the observed [TSH] signal, starting the charge period from 14.00 to 24.00 h, while the discharge period is observed from 24.00 to 14.00 h, etc. With the dashed line, we indicate the Fourier fundamental harmonic of the circadian signal.

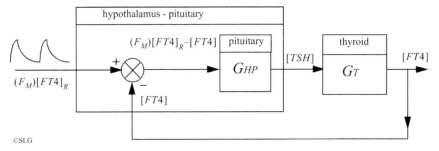

©SLG

Figure 16.5 Complete HPT feedback configuration with the circadian clock signal F_M modulating the internal reference $[FT4]_R$, resulting in the saw-tooth-like signal.

produce T4 and T3 hormones as a pre-charge operation preparing the energy demands to meet the metabolic processes for the next active period.

In Figure 16.5, the elementary feedback model of the homeostatic HPT regulation is depicted where the triangular charge–discharge signal modulates the internal reference $[FT4]_R$.

$$F_M[FT4]_R \tag{16.6}$$

According to the description and analysis of the closed-loop HPT of Chapter 9, we use the expression for the relationship between $[FT4]_R$ and the output $[FT4]$.

With the loop gain defined as

$$G_L = |G_{HP}G_T|, \tag{16.7}$$

we write

$$[FT4] = \frac{G_L}{1 + G_L}[FT4]_R \tag{16.8}$$

This implies that before $[FT4]$ increases, the $[TSH]$ has to increase first. Because the fast internal response of $[FT4]$ to a $[TSH]$ stimulation can be considered as a response in the open-loop situation, and that the inhibitory effect of $[FT4]$ on $[TSH]$ had not sufficient time to take place, we can write for the short-term effect:

$$[FT4] = G_T[TSH] \tag{16.9}$$

From Equations (16.7) and (16.8), we derive:

$$[TSH] = \left(\frac{1}{G_T}\right)\frac{G_L}{1 + G_L}[FT4]_R \tag{16.10}$$

According to the feedback model of Figure 16.5, and the related dependence of the output described in formula (Equation 16.8), we infer that the FT4 output follows the modulated reference value $[FT4]_R$. Variations of $[FT4]_R$ are directly translated to proportional variations in [TSH] resulting in a thyroidal response to produce [FT4] and [FT3]. This implies that the circadian clock variations are modulated on the reference variable $[FT4]_R$ of which the effect is observable as a circadian [TSH] variation. A similar effect is observable in hypothyroid individuals treated with L-T4 to achieve euthyroid [TSH] values near the individual set point. Only in this case of open HPT loop, we will not observe the fast pulsatile [TSH] oscillations [5, 11], but a steady increase and decrease in the circadian [TSH] rhythm. In a normal closed-loop HPT system, the pulsatile [TSH] oscillations are stimulating the thyroid to produce T4 and T3 resulting in related delayed and distributed output injections in plasma.

16.4 [TSH] Oscillations

The pulsatility of [TSH] is theorized to be caused when an overshoot of thyroid hormone concentrations reaches the momentary reference feedback threshold, $[FT4]_R$. Then the TSH secretion is temporarily shut down until the distribution process of FT4 and FT3 results in sufficient low concentrations to allow the [TSH] level to rise again. The distribution decay is observed on a time scale of hours, much faster than the normal 7-day half-life decay of [FT4] [16,17]. This control mechanism results in an oscillating behavior of [TSH] and [FT4] of which only the [TSH] variations are observable in the plots of all healthy participants because of the short measurement interval of 10 min. In individuals with a hypothyroid condition, this oscillatory behavior will not be observable because of the lack of a feedback mechanism. In Figure 16.6, a part of the rising reference signal $[FT4]_R$, together with the corresponding [TSH] measurements and the inferred oscillations of [FT4] between 11.40 and 14.30 h, is depicted. Although [FT4] is presented only in time intervals of 1 h, we can infer the [FT4] variations based on the observed variations of [TSH].

16.5 Data Analysis

The examination of the data set derived from the 38 participants showed a distinction in three categories presented in Table 16.1 [10].

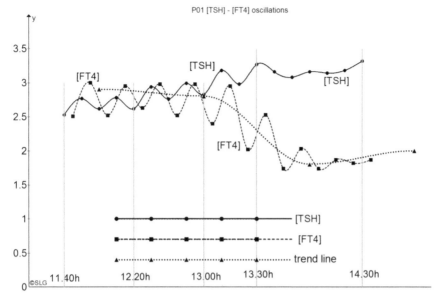

Figure 16.6 Measured oscillations of [TSH] (drawn) between 11.40 and 14.30 h with the inferred variations for [FT4] (dashed) following the originally depicted FT4 trend line (dotted) of P01 over the indicated interval.

Table 16.1 Three different categories of [TSH] ranges

Category I	Category II	Category III
Low [TSH]	$1.2 <$ [TSH] < 2 mU/L	High [TSH]
High [FT4] and [FT3]		Low [FT4] and [FT3]
$0.5 <$ [TSH] < 1.2 mU/L	Average [FT4] and [FT3]	$2 <$ [TSH] < 5 mU/L
$0.5 < \Delta$[TSH] < 1.8 mU/L		$1.8 < \Delta$[TSH] < 5.25 mU/L
$5 < G_L < 12$	$2.5 < G_L < 5$	$1 < G_L < 2.5$
$0.6 < \varphi < 1.35$	$0.35 < \varphi < 0.6$	$0.1 < \varphi < 0.35$
P# **9**, 18, 19, 21, 22, 25, 28, 29, 31, and 38	P# 2, 4, 5, 6, **7**, 10, 12, 15, 16, 17, 20, 23, 26, 27, 32, 34, and 37	P# 1, 3, 8, 11, 13, 14, 24, 30, 33, 35, and 36

The categories have been selected based on distinguishable set point parameters. A balanced division of the number of participants in each category: 25% in category I, 50% in category II, and 25% in category III, resulted in the selected boundaries of [TSH], [FT4], [FT3], φ, and G_L.

Based on the normal operating set point values for [FT4] and [TSH] of an individual, these parameters are derived from the 24-h average values of these variables [4]. The most prominent model parameters are the values for φ representing the sensitivity of the HP characteristic and the loop gain G_L determining the accuracy of the homeostatic control results and stability condition.

According to the first derivatives of the HP and thyroid model expressions, we find

$$G_L = |G_{HP}G_T| = \varphi A\alpha[TSH]\exp(-\alpha[TSH]) \qquad (16.11)$$

As an illustration of the differences in appearance, a belonging double 24-h plot of a suitable representative of the categories is presented in Figures 16.7–16.9. The results over the 24-h period are repeated for a better continuous overview. From Figures 16.7–16.9, we appreciate the circadian variations of TSH (drawn) and variations of the secretory patterns of [FT4] (dotted) and [FT3] (dashed).

The timing and delay properties of the appearance of [TSH] and [FT4] induce a local oscillatory behavior with limited amplitude [5, 11]. From a systems point of view, these stable oscillations are necessary for a proper long-term HPT system homeostasis [14]. Also the sinus forms of the fundamental harmonics are indicated for [TSH], [FT4] and [FT3].

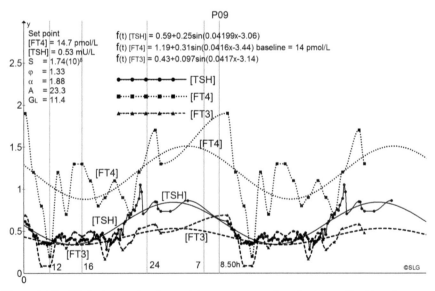

Figure 16.7 Clinical example of a participant representative of category I, with a characteristic low [TSH] (drawn), average, and relatively large variations in [FT4] (dotted).

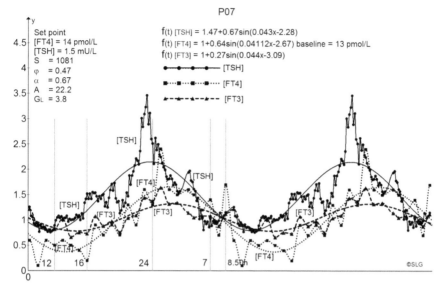

Figure 16.8 Clinical example of a representative participant under category II, with characteristic average [TSH] (drawn) and moderate variations in [FT4] (dotted). The correspondence between the variations in [FT3] (dashed) and [FT4] is evident.

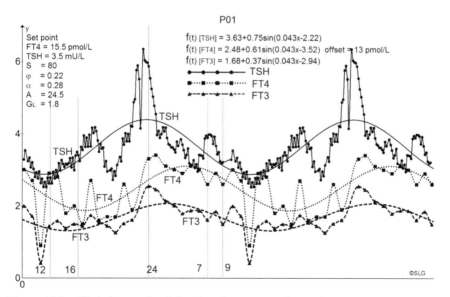

Figure 16.9 Clinical example of the circadian pattern of a participant with high average [TSH] (drawn) and moderate variations in [FT4] (dotted) with the corresponding variations of [FT3] (dashed).

Another oscillatory effect with longer periods is observed as a continuous rise of [TSH] over periods of about 1 h without a shutdown of the TSH production. In the example of P01 in Figure 16.9, we observe a continuous rise of [TSH] from 22.00 to 22.40 h. This implies that the rise of $[FT4]_R$ is faster than the rise of [FT4] due to thyroidal secretion. Between 22.50 and 23.00 h, we observe a steep decay of [TSH] implying a complete shutdown of the TSH secretion. This pattern of sustained rise and fall of [TSH] around midnight is commonly observed in the circadian variations. Another striking observation was the repetition of this phenomenon consistently after the first [TSH] peak after midnight. This peaking pattern, apart from the common 20-min repetition, can be interpreted as a second-order local instability with peaking amplitude varying from 0.5 to 1.5 mU/L.

Because the addition of T4 to the total base level will have an accumulative effect as a result of the relatively long half-life of T4, the pulsatile thyroid secretion results in an increasing base level of [T4] and [FT4], and to a lesser extent of [T3] and [FT3] owing to the shorter half-life of T3 [17]. The T4 cumulative effect is a continuous process and results normally in a homeostatic controlled condition of the closed-loop HPT. In a situation where hypothyroid individuals are on L-T4 replacement therapy, the titrated daily dose will keep the FT4 at the correct level. Conversion of [FT4] to [T3] and [FT3] by deiodinase and other metabolic effects results in a reduction of roughly 10% per 24 h of the pool of T4 [17]. The circadian base patterns of [TSH], [FT4], and [FT3] can be described with Fourier analysis resulting in the corresponding different fundamental sine-wave components.

In general, we observe a dominant fundamental harmonic for [TSH], but wave patterns with higher frequency components are recognizable as ultradian responses. This applies to both [FT4] and [FT3] variations. The circadian and ultradian variations are primarily caused by the accumulated results of [TSH] fluctuations and the corresponding thyroidal secretion responses of [FT4] and [FT3] which occur under local HPT delay, distribution, and feedback control. All these mentioned effects are observed in all possible varieties in individuals and share the basal triangular circadian relaxation signal of Figure 16.1 together with the corresponding lead reference of the [TSH] variations.

16.6 Fourier Analysis

A closer look at the inferred frequencies and phases of the fundamental harmonics reveals the presence of a 24-h period fundamental harmonic for

[TSH], [FT4], and [FT3]. Because the internal reference [FT4]$_R$ is modulated in a natural circadian mode, the resulting variations follow each other with [TSH] as the first leading signal. The response of the thyroid will follow after the [TSH] stimulus, resulting in the secretion of [T4] and [T3] simultaneously. The fundamental harmonics of [FT4] and [FT3] have more or less the same frequency as [TSH] but show a wide variety of effective phase shifts compared to the phase reference with [TSH]. The individual metabolic characteristics and physiological properties of delay and distribution can be observed in the individual data analyses of all participants. With the aid of some fitting functions using a graph-plotting software, we can visualize the included basic circadian sinusoidal variations, but also the higher harmonics of which the circadian signals consist.

The general expression of such a harmonic series can be noted as

$$g(t) = h_0 + h_1 \sin(\omega t + \phi_1) + h_2 \sin(2\omega t + \phi_2) + \ldots + h_N \sin(N\omega t + \phi_N),$$
$$(16.12)$$

resulting in the general formulation:

$$g(t) = \sum_{n=0}^{n=N} h_n \sin(n\omega t + \phi_n), \qquad (16.13)$$

and applying the compound angle formula

$$\sin(n\omega t + \phi_n) = \sin(n\omega t)\cos(\phi_n) + \cos(n\omega t)\sin(\phi_n), \qquad (16.14)$$

we find

$$g(t) = a_0 + a_1 \cos(\omega t) + a_2 \cos(2\omega t) + \ldots a_N \cos(N\omega t) + \\ b_1 \sin(\omega t) + b_2 \sin(2\omega t) + \ldots b_N \sin(N\omega t) + \qquad (16.15)$$

of which all coefficients, *a* and *b,* are real.

The function $g(t)$ is periodic with period $T = \frac{2\pi}{\omega}$.

From Equation (16.15), it is obvious that periodic functions like $g(t)$ can be subdivided in one ore more sine or cosine functions with different amplitudes and frequencies. Using graphical software available freely (*Graph 4.4*), the following generalized fitting function was used to find the ground harmonic and other ultradian variations:

$$g(t) = \$a + \$b * \sin(\$c * x + \$d)$$

The term $c can be chosen such that the lowest possible harmonic value for *x* can be found. In our examples, $0.02 < \$c < 0.06$ was being experimented to find the most suitable results. Similarly, the ultradian harmonics can be found for higher values of $c.

For the coefficients a_n and b_n, we find the following expressions:

$$a_n = \frac{2}{T} \int_0^T f(t) \cos(n\omega t)dt \qquad (16.16)$$

and

$$b_n = \frac{2}{T} \int_0^T f(t) \sin(n\omega t)dt \qquad (16.17)$$

In Equations (16.16) and (16.17), *f(t)* represents the function of which these coefficients can be determined.

The following analysis shows an example using the triangular waveform function of Figure 16.10.

From the symmetry properties, we can derive the function consisting of cosine terms of odd harmonics. The first half period of *f(t)* is a linear function with a linear coefficient 4A/T, and a zero for *t* = T/4. Then we find:

$$f(t) = \frac{4A}{T} \left(t - \frac{T}{4} \right) \qquad (16.18)$$

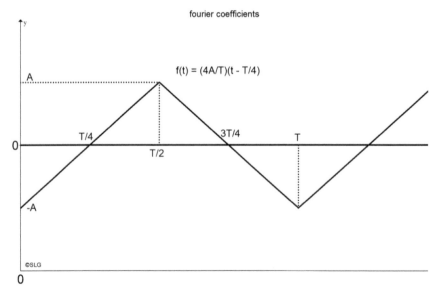

Figure 16.10   A symmetric linear periodic function with both symmetric and translational symmetry.

For mirror symmetric functions, we can derive that:

$$a_n = \frac{4}{T} \int_0^{T/2} \left\{ \frac{4A}{T} \left(t - \frac{T}{4} \right) \cos(n\omega t) \right\} dt, \qquad (16.19)$$

resulting in:

$$a_n = \frac{16A}{T^2} \left[\left(\frac{t - T/4}{n\omega} \right) \sin(n\omega t) + \left(\frac{1}{n^2 \omega^2} \right) \cos(n\omega t) \right]_0^{T/2} = \frac{-8A}{n^2 \pi^2}, \qquad (16.20)$$

with odd values for n, or $n = 1, 3. 5, \ldots$

16.7 Convolution and Deconvolution

Reflecting back on the principles of system and control theory, every system element in biological systems can be defined as linear time invariant (LTI). Time invariance means that the output response $y(t)$ on a certain input $x(t)$ is identical with a certain time delay T. Therefore, $Ay(t - T) = x(t - T)$. This implies that any system part defined as a single signal transfer block as discussed in Chapter 5 is characterized by the impulse response $h(t)$.

The output of a time-invariant system block is then defined as the convolution of the input signal $x(t)$ and $h(t)$ noted as $y(t) = h(t) * x(t)$.

The principle of deconvolution is defined as a signal processing operation to deconvolute a recorded measurement result h and is written as:

$$f * g = h \qquad (16.21)$$

In general, we want to remove the interfering signal g from our measurements h to restore the original signal f. In some cases, g is deterministic and known. In other situations, g is unknown and has to be estimated. Besides the interfering signal g, we also encounter an added noise component ε which we can note as:

$$(f * g) + \varepsilon = h \qquad (16.22)$$

When the properties of g are unknown, and possibly have to be estimated and we have a large contribution of noise (ε), it is virtually impossible to perform a proper deconvolution operation. Normally, the deconvolution is found with the Fourier integral transform of the components f, g, and h according to:

$$F(\omega) = \int_{-\infty}^{\infty} f(\tau) \exp(-j\omega\tau) d\tau, \qquad (16.23)$$

resulting in transformation of the respective functions from time domain to the frequency domain:

$$f(t) --> F(\omega) g(t) --> G(\omega) \text{ and } h(t) --> H(\omega) \qquad (16.24)$$

In the absence of noise, we find:

$$f * g = h --> F.G = H, \qquad (16.25)$$

from which we derive that the original signal (as Fourier transformed function) is found according to:

$$F = \frac{H}{G} \qquad (16.26)$$

16.8 Thyroid Response

From Table 16.1, we can indicate the categories I–III as distinction in the responsivity of the thyroid to the stimulating variations of [TSH]. For this purpose, we define the circadian thyroid responsivity as:

$$G_{T24h} = \frac{\text{Circadian amplitude } [FT4]}{\text{Circadian amplitude } [TSH]} \qquad (16.27)$$

The thyroid responsivity represents a circadian figure of merit.

From the average value of [TSH] and [FT4] over 24 h, we define the average set point for the participant. Based on this set point value, we can follow the set point displacement during 24 h to find the position on the thyroid characteristic. This set point position defines the momentary or dynamic thyroid responsiveness on variations of [TSH]. This effect is easily observable from the participants with low average [TSH] resulting in relatively high variations in [FT4] and vice versa.

16.9 Set Point Stability

During the movement of the HPT set point position as we follow the variations in concentration of [FT4] and [TSH], we encounter at every different set of [FT4]–[TSH] values a new set of belonging HP and thyroid curves. According to an example of possible variations, we will use the fundamental [FT4] and [TSH] harmonics of three cases in the sample to find the values as presented in Table 16.2.

Table 16.2 Overview of the different thyroid responses

	$P_{0(average)}$		P_1		P_2		P_3		P_4	
P	[FT4]	[TSH]	[FT4]	[TSH]	[FT4]	[TSH]	[FT4]	[TSH]	[FT4]	[TSH]
	av	Av		max	2	min	max	3	min	4
09	15.07	0.5	15.2	0.89	14.2	0.35	15.3	0.738	14.7	0.71
07	14.02	1.48	14.6	3.09	13.6	0.81	15.4	2.24	13.1	0.92
01	15.5	3.5	15.6	4.4	15.3	2.87	16.1	3.9	14.9	3.34

When we look at the example of one such case (P01) in Figure 16.8, we appreciate the maximum [TSH] value at about 24 h as [TSH] = 4.4 mU/L during which the value of [FT4] = 2.76 + the offset value.

The offset value or baseline correction amounts to 13 pmol/L, whereupon we find [FT4] = 15.76 pmol/L. When we calculate the belonging thyroid characteristic, we find A = 24.9 and α = 0.23. A similar calculation for the minimum value of [TSH] fundamental gives [TSH] = 2.87 mU/L at about 11.00 h while [FT4] = 2.3 + 15 = 15.3 pmol/L. This set point value results in belonging thyroid parameters of A = 24.2 and α = 0.35. This example shows that the secretion maximum A is relatively constant but the steepness factor α is distinctly different as depicted in Figure 16.11. In Figure 16.12 we depict the different positions of the set point over 24 h.

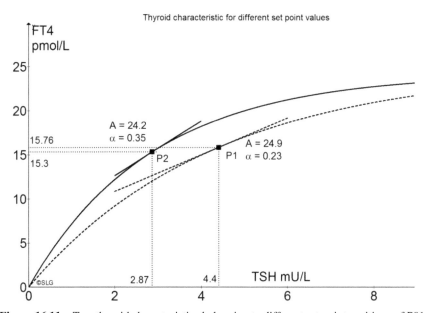

Figure 16.11 Two thyroid characteristics belonging to different set point positions of P01.

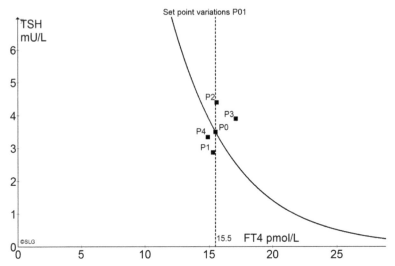

Figure 16.12   Different set point positions P_0 to P_4 depending on the time of day derived as example from P01 indicated in Table 16.2. The difference between the highest value of [FT4] indicated in P_3, compared to the lowest [FT4] in P_4 is smaller than 10%, as has to be expected.

When we calculate the steepness in P1 and P2, we find according to Equation (16.27):

$$S_{P1} = A\alpha \exp(-\alpha[TSH]) = (24.2).(0.35) \exp(-0.35(4.4))$$
$$= 1.8 \text{ pmol/mU}$$

and

$$S_{P2} = A\alpha \exp(-\alpha[TSH]) = (24.9).(0.23) \exp(-0.23(2.87))$$
$$= 2.9 \text{ pmol/mU}$$

This result shows a difference in dynamic thyroid secretory performance. From the previous analysis, we can conclude that the circadian modulation of the internal set point reference $[FT4]_R$ will have consequences for the set point position. This results in a total of five different positions as indicated in Figure 16.9. The 24-h average value of FT4 = 15.5 pmol/L which shows little deviation.

16.10 Discussion

Circadian rhythms of various hormones have been investigated for some decades [5–7]. The improvement of measurement technology determining the concentration values of these hormones made it possible to analyze 24-h hormone concentration time series with improved accuracy. For example, today we have the possibility for a continuous glucose measurement that can be applied for diabetes patients [18]. Such a continuous measurement and display feature may not be available for [TSH], [FT4], and [FT3] as yet though the technology of hormonal assays continues to improve in detection sensitivity and accuracy.

In this, chapter we studied the circadian variations of 38 healthy participants from whom we could determine a distinct pattern in the variations of [TSH], [FT4], and [FT3]. Together with the parameter method from earlier published HPT system papers [3, 4, 19], we managed to use individual set point parameter values to find methods for categorization of investigated participants. The origin of the circadian characteristic pattern of [TSH] is complex. We theorize for the first time that its underlying biological mechanism could ultimately be deciphered based on our understanding in terms of electrodynamics theory as a process induced by a modulating charge–discharge signal of the internal reference $[FT4]_R$ for the HPT set point as depicted in Figure 16.5. We speculate that such diurnal variation has the function to prepare the individual for the next active period with a "pre-charge of T4" during the evening. Hypothyroid persons on a daily L-T4 substitution therapy show a similar charge–discharge effect of [FT4] as induced by the circadian [FT4] charge–discharge characteristic because of their daily intake of L-T4 medication.

A general signal analysis and description was earlier published by Beersma et al. in 2005 [12]. On the charge–discharge signal, we find the [TSH] drive following the drive on $[FT4]_R$ resulting in a superposition of TSH with relative small short time variations and an amplitude of about 0.4 mU/L as a result of the feedback effects of [FT4]. These oscillations should not be marked as random variations or noise in the [TSH] data because the sensitivity and accuracy of the assay will rule out such an assumption. Similar variations in [FT4] and [FT3] were undetectable over measurement intervals of one hour but were inferred from the [TSH] variations detectable via high-resolution 10-min measurement intervals.

In the total circadian effect, we find a fundamental harmonic of the [TSH] variations behaving as the leading or initiating signal for the later occurring harmonics of [FT4] and [FT3]. In all cases, we find a phase lag for [FT3] relative to the phase of [TSH]. Depending on the individual temporal deiodinase activity, we generally find [FT4] lagging after [FT3]. Because the set point fluctuates to the same rhythm and amplitude as the fundamental harmonic for [TSH] and [FT4], we can distinguish different properties of the actual secretory dynamics of the thyroid. In this way, we can define the circadian figure of merit (i.e., performance or efficiency) for the thyroid which represents the thyroidal responsiveness over 24 h as indicated in Equation (16.27). The overall system stability is mainly defined by the loop gain factor G_L of the HPT system which has to be greater than unity.

16.11 Concluding Remarks

A circadian scan of a person can provide a detailed temporal profile of the [TSH], [FT4], and [FT3] dynamics. Together with individual HPT system parameters, we can develop a clear understanding of normal distinctive differences and possible pathologic anomalies. We distinguished three categories of HPT system behavior which could be helpful in diagnostics.

Furthermore, we theorized that the circadian pattern of [TSH] is a result of the physiological need to recharge the HPT system with a sufficient amount of [T4] to prepare the body for the next active period. In healthy persons, this results in a measurable pulsatile behavior of [TSH] which in turn should induce similar pulsatile effects on the secretory behavior of [FT4] and [FT3]. It is a limitation of the current study that these could not be measured because of the 1-h measurement periods. Looking at the [TSH] variations in the different categories as indicated in Table 16.1, we have a confirmation that the [TSH] HP output response fully conforms to the HP characteristic. This hypothesis could be confirmed by similar [TSH] measurements performed in hypothyroid persons from whom we could expect an open HPT loop in which no feedback effect between [TSH] and [FT4] will be expected ($G_L < 1$). Based on the feedback model of the HPT system of Figure 16.5, we expect similar [TSH] variations in persons who have no functioning thyroid but without the pulsatile effects of [TSH].

References

[1] Colucci, P., Seng Yue, C., Ducharme, M., and Benvenga, S. (2013). A review of the pharmacokinetics of levothyroxine for the treatment of hypothyroidism. *Eur. Endocrinol.* 9, 40–47.
[2] Bianco, A. C., Salvatore, D., Gereben, B., Berry, M. J., and Larsen, P. R. (2002). Biochemistry, cellular and molecular biology, and physiological roles of the iodothyronine selenodeiodinases. *Endocr. Rev.* 23, 38–89.

[3] Goede, S. L., Leow, M. K., Smit, J. W. A., Klein, H. H., and Dietrich, J. W. (2014). Hypothalamus-pituitary-thyroid feedback control: implications of mathematical modeling and consequences for thyrotropin (TSH) and free thyroxine (FT4) reference ranges. *Bull. Math. Biol.* 76, 1270–1287. doi: 10.1007/s11538-014-9955-5

[4] Leow, M. K., and Goede, S. L. (2014). The homeostatic set point of the hypothalamus-pituitary-thyroid axis–maximum curvature theory for personalized euthyroid targets. *Theor. Biol. Med. Model.* **11**:35. doi:10.1186/1742-4682-11-35

[5] Veldhuis, J. D., Keenan, D. M., and Pincus, S. M. (2008). Motivations and methods for analyzing pulsatile hormone secretion. *Endocr. Rev.* 29, 823–864. doi: 10.1210/er.2008-0005

[6] Roelfsema, F., Pereira, A. M., Adriaanse, R., Endert, E., Fliers, E., Romijn, J. A., et al. (2010). Thyrotropin secretion in mild and severe primary hypothyroidism is distinguished by amplified burst mass and basal secretion with increased spikiness and approximate entropy. *J. Clin. Endocrinol. Metab.* 95, 928–934.

[7] Roelfsema, F., and Veldhuis, J. D. (2013). Thyrotropin secretion patterns in health and disease. *Endocr. Rev.* 34, 619–657.

[8] Saravanan, P., Siddique, H., Simmons, D. J., Greenwood, R., and Dayan, C. M. (2007). Twenty-four hour hormone profiles of TSH, free T3 and free T4 in hypothyroid patients on combined T3/T4 therapy. *Exp. Clin. Endocrinol. Diabetes* 115, 261–267.

[9] Lechan, R. M., and Fekete, C. (2004). Feedback regulation of thyrotropin-releasing hormone (TRH): mechanisms for the non-thyroidal illness syndrome. *J. Endocrinol. Invest.* 27(Suppl. 6), 105–119.

[10] Jansen, S. W., Roelfsema, F., van der Spoel, E., Akintola, A. A., Postmus, I., Ballieux, B. E., et al. (2015). Familial longevity is associated with higher TSH secretion and strong TSH-fT3 relationship. *J. Clin. Endocrinol. Metab.* 100, 3806–3813. doi: 10.1210/jc.2015-2624

[11] Li, G., Liu, B., and Liu, Y. (1995). A dynamical model of the pulsatile secretion of the hypothalamo-pituitary-thyroid axis. *Biosystems* 35, 83–92.

[12] Beersma, D. (2005) Why and how do we model circadian rhythms? *J. Biol. Rhythms* 20, 204–313. doi: 10.1177/0748730405277388

[13] Olde Scheper, T., Klinkenberg, D., Pennat, C. J., and van Pelt, J. (1999). A mathematical model for the intracellular circadian rhythm generator. *J. Neurosci.* 19, 40–47.

[14] Astrom, K. J., and Murray, R. M. (2009). *Feedback Systems: An Introduction for Scientists and Engineers*, 2nd Edn. New Jersey, NJ: Princeton University Press.

[15] Persani, L., Terzolo, M., Asteria, C., Orlandi, F., Angeli, A., and Beck-Peccoz, P. (1995). Circadian variations of thyrotropin bioactivity in normal subjects and patients with primary hypothyroidism. *J. Clin. Endocrinol. Metab.* 80, 2722–2728. doi: 10.1210/jcem.80.9.7673415

[16] Jonklaas, J., Burman, K. D., Wang, H., and Latham, K. (2015). Single-dose T3 administration: kinetics and effects on biochemical and physiological parameters. *Ther. Drug. Monit.* 37, 110–118.

[17] Bauer, L. A. (2008). *Applied Pharmacokinetics*, 2nd Edn. New York, NY: McGraw Hill. DOI: 10.1036/0071476288

[18] Akintola, A. A., Noordam, R., Jansen, S. W., de Craen, A. J., Ballieux, B. E., Cobbaert, C. M., et al. (2015). Accuracy of continuous glucose monitoring measurements in normo-glycemic individuals. PLOS ONE 10:e0139973. doi:10.1371/journal.pone.0139973

[19] Goede, S. L., Leow, M. K., Smit, J. W. A., and Dietrich, J. W. (2014). A novel minimal mathematical model of the hypothalamus-pituitary-thyroid axis validated for individualized clinical applications. *Math. Biosci.* 249, 1–7. doi: 10.1016/j.mbs.2014.01.001

17

HPT Simulation with Trans-linear Circuits

"Learning by doing, peer-to-peer teaching and computer simulation are all part of the same equation."

–Nicolas Negroponte (1943–Present)

17.1 Introduction

Until this stage, we described and discussed various ideas surrounding the physiology and mathematical constructs of the HPT system. All these results can be verified by actual measurements in real patients. Such mathematical models clearly find practical applications at the bedside. However, mathematical models are useful even under circumstances when ordinary experiments would be impossible or unethical to execute on real humans because the alteration of conditions and/or parameters can prove life-threatening with fatal consequences. These impediments can be circumvented by the use of models that mimic the real physiological properties of the subject of interest so as to allow simulations. A first glance on the methods already in use in biologic simulation is given in the book of Kai Velten [1] and an appropriate paper by Eisenberg et al. [2]. All other books on mathematical biology, modeling, and simulation provide only the generalized principles of mathematical modeling but lack the specifics we discuss in this chapter.

In previous chapters about physiological modeling, we successfully applied simulation techniques based on the behavioral characteristics of passive electrical networks consisting of resistors and capacitors. Physiology apparently lends itself well to modeling methods involving electrical network because the network element behavior is directly translatable to the underlying physiologic behavior [3, 4]. In general, before using any form of simulation, it is of utmost importance to characterize the physiology of interest and to validate the modeling result with real data of the subject under

255

investigation. We have to keep in mind that a generalization of physiologic behavior of a group of subjects is fundamentally impossible and therefore not applicable for simulation purposes. A nice example can be found in Chapter 14.

This approach is fundamentally different compared to curve-fitting methods commonly applied in biologic modeling which has no relationship with the underlying physiology or any other mechanism. Previously, we discussed the quasi-static behavior of the HP and thyroid. Here, we encounter a mathematical relationship that is not dynamic but parametric with input and output variables that are time-independent. Therefore, as a first modeling step, we have to introduce relationships similar to the properties of active network elements like vacuum tubes, field-effect transistors, and in our case, bipolar transistors.

With these electrical network models, we can describe exponential functions, quadratic functions, and all kinds of variants based on the properties of the active transistor elements. This modeling approach and description is known as trans-linear circuit analysis [5, 6]. Based on the previously described relationships between [FT4] and [TSH] of the HP and thyroid, we can model exactly the same mathematical representations with trans-linear circuits. With these circuits, we can arrange the HPT feedback loop exactly as previously described and have the possibility to study a spectrum of behaviors under an extensive range of conditions without the burden of these experiments on real people. In the following, we will first discuss the closed HPT feedback loop based on the large-signal behavior of HP and thyroid. This configuration will be synthesized with trans-linear circuits and analog electronic circuits.

A common structure consisting of a thyroid gland secreting thyroid hormones and a controlling mechanism to maintain a homeostatic balance can be practically found in all vertebrates. Thyroid hormone levels are maintained by means of negative feedback control in every healthy person. Hormone secretion generally does not blow up uncontrollably. This implies that the thyroid gland when stimulated by a control variable like thyroid stimulation hormone (TSH) will never show an exponential increase. On the contrary, the gland will be characterized by saturation behavior in the largest range of stimulation. This common behavior also applies for inhibitory behavior, representing the inverse action of stimulation or inhibition. Nevertheless, both stimulation and inhibition are strongly related.

The HP system has been described as non-linear whereas the thyroid as a controlled secretory component [8–12]. The validity of these models has been demonstrated. Still, there is the need to investigate the real-time endocrine behavior in a laboratory environment. For this purpose, we will present an electronic functional equivalent, translatable to the underlying physiologic reality. One of the purposes of this book is dedicated to cooperation between various disciplines. At first, we have the recognition from clinical experts in endocrinology who will be able to compare the simulator behavior with practical situations. On the other hand, we need the expertise of electronic engineers to model the physiologic behavior and implement the various system behavioral components and the related measurements in electronic circuits. Both disciplines should try to understand each other and work together to realize the synergy of both knowledge realms and come to an integrated understanding incorporated by the proposed simulator. The background of this work is based on knowledge and experience in endocrinology. Together with electronic simulation possibilities, this combination will realize a reality check without the burden on animals or humans.

17.2 Generalized Stimulated Gland Characteristics

As mentioned in the introduction, a stimulated gland can have three possible characteristics as depicted in Figure 17.1.

For a natural gland, we encounter only the exponential saturating response because of the limited secretory capacity. The general form is presented in Figure 17.2 [11, 12].

The function in Figure 17.2 is defined as

$$y = S\left(1 - \exp(-\varphi x)\right), \tag{17.1}$$

with saturation level A and the steepness defined as

$$\frac{dy}{dx} = \varphi S\left(\exp(-\varphi x)\right) \tag{17.2}$$

In $x = 0$, we find the largest value for the steepness

$$\left[\frac{dy}{dx}\right]_{x=0} = \varphi S \tag{17.3}$$

Besides the property of exponential secretory saturation to an asymptotic level A, this function can also be interpreted in the inverse sense, known as inhibition. The inhibitory action is then expressed as minimum inhibition at $x = 0$ and full inhibition when $f(x)$ reaches A.

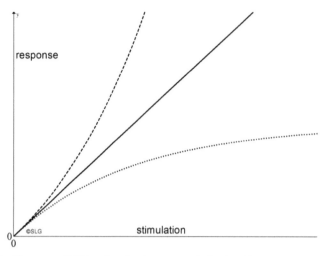

Figure 17.1   Three possibilities for gland secretory behavior. The solid line represents the linear response, the dashed line represents an exponential response, and the dotted line represents an exponential saturating response.

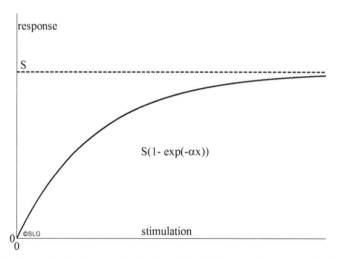

Figure 17.2   Saturating behavior of a stimulated gland. The maximum secretion is indicated with level *S*.

In such a case, we can write this new function as

$$f_{in}(x) = S - y = S - S\left(1 - \exp(-\varphi x)\right) \qquad (17.4)$$

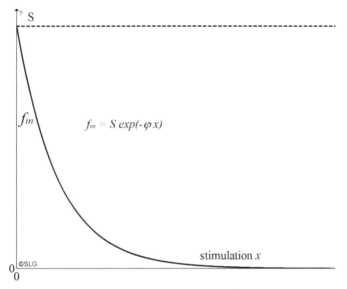

Figure 17.3   Inverted stimulation by means of induced inhibition.

Then we find the inhibitory function as

$$f_{in}(x) = S\exp(-\varphi x) \tag{17.5}$$

This expression is valid because for $x = 0$, we find

$$f_{in}(x) = S, \tag{17.6}$$

as depicted in Figure 17.3.

This form of stimulated inhibition is a well-known TSH response of the HP system on the detection of circulating concentrations free T4 or [FT4]. A generalized form of these processes is commonly described as

$$\frac{dy}{dx} = Ay \tag{17.7}$$

The common solution of this DV is written as

$$\ln(y) = Ax + B \tag{17.8}$$

This expression is normally used to depict the log-linear relationship of [TSH] and [FT4] as represented by Figure 17.4.

From Figure 17.4, we appreciate that the steepness of the function is represented by the coefficient A and a shifting parameter, B. In clinical data

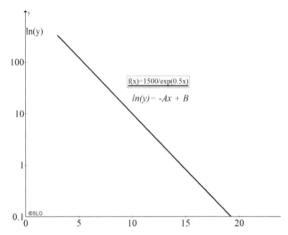

Figure 17.4   Logarithmic vertical scale and linear horizontal scale for $\ln(y) = Ax + B$.

representations, Figure 17.4 is the standard to depict the log-linear relationship. With the following transform, we can find the exponential linear–linear expression:

$$\exp\left(\ln(y)\right) = \exp\left(Ax + B\right) = y = \exp(Ax).\exp(B) \qquad (17.9)$$

The value of $\exp(B)$ represents here the shifting parameter S and the coefficient A represents the exponential coefficient φ. This results in the expression found in Equation (17.5). Both functions: the saturation and the stimulated inhibitory one can be realized with electronic components using the property of an exponential transfer function.

In the following section, the general closed-loop feedback configuration of the HPT system will be discussed.

17.3 Large-signal Intercept Set Point

Previously, we analyzed the closed HPT feedback loop with linearized transfer function blocks where the signal transfer was defined around a fixed point of operation together with a reference value [FT4]$_R$ for [FT4] [11]. The feedback loop can also be presented as the physiological loop in which we will not indicate the reference value [FT4]$_R$ as was used in the previous feedback block diagrams. The internal reference value is now factored based on the chosen value of the set point with [FT4] = 15 pmol/L and [TSH] = 1.5 mU/L. In Figure 17.5, the block diagram of the HPT physiology is

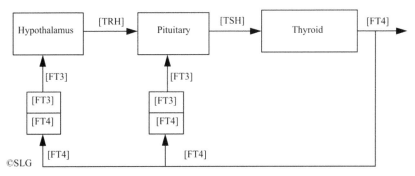

Figure 17.5 Physiological HPT feedback loop presentation.

depicted and the internal reference value is derived from the point of maximum curvature of the HP curve.

From Figure 17.5, we can find the equations describing the equilibrium condition for the negative HPT feedback operation.

For the HP characteristic, we found

$$[TSH] = S \exp(-\varphi[FT4]), \qquad (17.10)$$

and the transfer characteristic of the thyroid

$$[FT4] = A\{1 - \exp(-\alpha[TSH])\} \qquad (17.11)$$

These equations are interacting under the condition that Equation (17.10) is substituted into Equation (17.11) resulting in a sole implicit equation for [FT4]

$$[FT4] = A\{1 - \exp(-\alpha S \exp(-\varphi[FT4]))\} \qquad (17.12)$$

When we substitute

$$[FT4] = x, \qquad (17.13)$$

we can solve the following equation

$$x = A\{1 - \exp(-\alpha S \exp(-\varphi x))\} \qquad (17.14)$$

Equation (17.14) is an implicit function of x of which we will find the solution by the following operation

$$F_1 : y = x \qquad (17.15)$$

and

$$F_2 : y = A\{1 - \exp(-\alpha S \exp(-\varphi x))\} \qquad (17.16)$$

The solution for the set point is found as the intercept point of F_1 and F_2 according to

$$F_1 = F_2 \qquad (17.17)$$

The proposed set point example is based on [FT4] = 15 pmol/L and [TSH] = 1.5 mU/L. When we use the earlier derived set point equations, we have:

$$\varphi = \frac{1}{[TSH]_{sp}\sqrt{2}} \qquad (17.18)$$

$$S = [TSH]_{sp} \exp(\varphi[FT4]_{sp}) \qquad (17.19)$$

$$A = \frac{[FT4]_{sp}}{0.632} \qquad (17.20)$$

$$\alpha = \frac{1}{[TSH]_{sp}} \qquad (17.21)$$

Then, we find, respectively, φ = 0.47, S = 1766, A = 23.7, and α = 0.67 resulting in the set point equation:

$$x = 23.7\{1 - \exp(-1183\exp(-0.47x))\} \qquad (17.22)$$

The intersection of Equations (17.14) and (17.15) can be present as the solution result depicted in Figure 17.6.

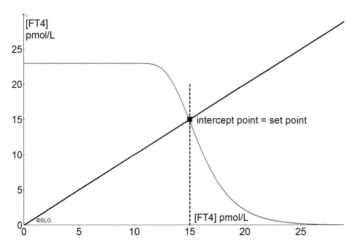

Figure 17.6 Intercept point of the large-signal representation to find [FT4] of the physiological HPT loop.

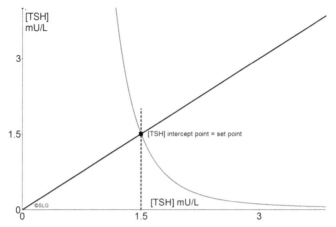

Figure 17.7 [TSH] Intercept point of the large-signal representation of the physiological HPT loop based on the set point parameters belonging to [FT4] = 15 pmol/L and [TSH] = 1.5 mU/l.

A similar expression for [TSH] can be found when we substitute Equation (17.11) into Equation (17.10).
This results in

$$[TSH] = S \exp(-\varphi A\{1 - \exp(-\alpha[TSH])\}), \tag{17.23}$$

and with the set point parameters $\varphi = 0.47$, $S = 1766$, $A = 23.7$, and $\alpha = 0.67$.
Upon substituting [TSH] = x, we find

$$x = 1766 \exp(-11.14 + 11.14 \exp(-0.67x)), \tag{17.24}$$

resulting in the graph of Figure 17.7.

17.4 Differential Amplifiers

The exponential behavior of the hypothalamus and the thyroid can exactly be implemented with the use of bipolar transistors in a trans-linear electronic circuit. These circuit blocks behave then exactly as the modeled HP and thyroid. This opens the possibility to realize these functions in hardware with which we can study the behavior in real time. We build a simulator with hardware in such a way that the model behavior is part of the design implementation. Before we present these functional implementations, some electronic and trans-linear principles will be discussed.

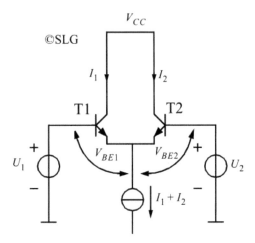

Figure 17.8   Basic differential pair.

In the following section, we will introduce non-linear electronic circuits and the related non-linear properties of bipolar transistors. One of the most common amplifier and comparator configurations is represented in the form of a differential transistor pair. This basic configuration is depicted in Figure 17.8.

For the large-signal analysis of the configuration of Figure 17.8, we use the following expressions.

$$I_{E1} = I_s \exp \left(\frac{V_{BE1}}{V_T} \right),$$

(17.25)

with

$$V_T = \frac{kT}{q}$$

(17.26)

where

k represents Boltzmann's constant.
T represents the absolute temperature in kelvin.
q represents the elementary electrical charge.
I_s represents the temperature dependent junction leakage current and is defined here with $T = 300\text{K}$.

Then $V_T = 26$ mV

$$\frac{I_{E1}}{I_{E2}} = \frac{I_s \exp\left(\frac{V_{BE1}}{V_T}\right)}{I_s \exp\left(\frac{V_{BE2}}{V_T}\right)} = \exp\left(\frac{V_{BE1} - V_{BE2}}{V_T}\right) = \exp\left(\frac{V_D}{V_T}\right) \quad (17.27)$$

$$I_{E1} + I_{E2} = 2I \quad (17.28)$$

From Equation (17.27), we find

$$I_{E1} = I_{E2} \exp\left(\frac{V_D}{V_T}\right) \quad (17.29)$$

When we write:

$$\frac{V_D}{V_T} = V_N, \quad (17.30)$$

then

$$I_{E2} \exp(V_N) + I_{E2} = 2I \quad (17.31)$$

$$I_{E2}\{\exp(V_N) + 1\} = 2I \quad (17.32)$$

$$I_{E2} = \frac{2I}{1 + \exp(V_N)}, \quad (17.33)$$

and from

$$I_{E2} = \frac{I_{E1}}{\exp(V_N)} = I_{E1} \exp(-V_N), \quad (17.34)$$

it follows that

$$I_{E1} = \frac{2I}{1 + \exp(-V_N)} \quad (17.35)$$

We present the differential input voltage V_N as referenced to volt as 26 mV at a junction temperature of $300°$K. This implies that complete current switching is performed at $V_N = 130$ mV. The normalized input voltage and output current results are depicted in Figure 17.9.

The linear region of operation is defined as $-0.5 < V_N < 0.5$; this region is normally used for linear signal processing operations. The input voltage range can be extended by the application of so-called emitter degeneration. This means that resistors are placed in series with the emitter. This configuration is depicted in Figure 17.10.

The input output transfer characteristics are presented in Figure 17.11 for different values of R_E.

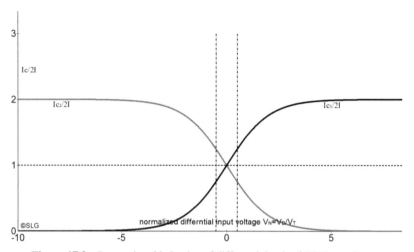

Figure 17.9   Large-signal behavior of differential pair of NPN transistors.

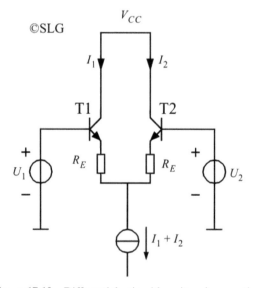

Figure 17.10   Differential pair with emitter degeneration.

When we neglect the influence of the base current, the collector current with emitter degeneration is represented by

$$I_c = \frac{g_m V_D}{1 + g_m R_E} \tag{17.36}$$

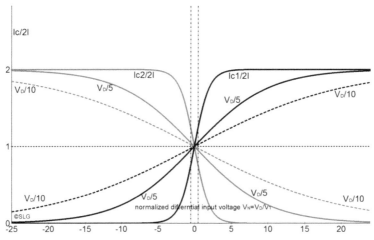

Figure 17.11 Normalized transfer characteristics with emitter degeneration.

This implies that the differential input signal will be attenuated by the R_E degeneration.

The amount of degeneration is indicated by the attenuation factor α, with $\alpha = \frac{1}{R_E}$.

The input signal variable V_N is now replaced by $\frac{V_N}{R_E}$ resulting in

$$I_{E2} = \frac{2I}{1 + \exp(\alpha V_N)} \tag{17.37}$$

and

$$I_{E1} = \frac{2I}{1 + \exp(-\alpha V_N)} \tag{17.38}$$

The normalized transfer characteristics are depicted in Figure 17.11.

From Figure 17.11, we appreciate the extended range for linear signal transfer which will increase by increasing values of the emitter series resistor R_E.

17.5 Trans-linear Circuits

17.5.1 Introduction

Trans-linear circuits form a class of circuits with which non-linear signal processing can be achieved. We will discuss a a simple example in the form of a current mirror as depicted in Figure 17.12. The signal operation of

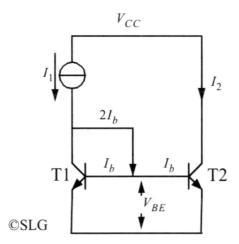

Figure 17.12   Elementary trans-linear circuit as the current mirror.

trans-linear circuits is based on the exponential relationship of the voltage over a P–N junction and the related current.

In Figure 17.12, the collector currents are determined by the value of the common base–emitter voltage V_{BE} where the value of the base currents is neglected. The driving current I_1 is mirrored in the mirror transistor T2 such that $I_1 = I_2$.

The basic equation that describes bipolar trans-linear circuits is

$$I_C = I_s \exp\left(\frac{qV_{BE}}{kT}\right), \tag{17.39}$$

where

I_s represents the thermal leakage current.
q represents the elementary value of charge.
T is the absolute temperature.
k represents Boltzmann's constant.

In the following equations, we write

$$\frac{kT}{q} = V_T, \tag{17.40}$$

where for values of $T = 300\,K$ (20°C), we find

$$V_T = 26\,mV \tag{17.41}$$

Based on these equations, we can synthesize various forms of non-linear signal processing.

17.5.2 Examples of Signal Processing with Trans-linear Circuits

Based on the principles of the current mirror, we can elaborate further on several properties of a trans-linear loop. This idea is depicted in Figure 17.13. From Figure 17.13, we can derive the following equations.

$$V_{BE1} = V_{BE2} + V_x, \tag{17.42}$$

or

$$V_x = V_{BE1} - V_{BE2} \tag{17.43}$$

From Equation (17.39), we derive:

$$\exp\left(\frac{qV_{BE}}{kT}\right) = \frac{I_C}{I_s} \tag{17.44}$$

$$\frac{qV_{BE}}{kT} = \ln\left(\frac{I_C}{I_s}\right) \tag{17.45}$$

$$V_T = \frac{q}{kT} \tag{17.46}$$

$$V_{BE} = V_T \ln\left(\frac{I_C}{I_s}\right) \tag{17.47}$$

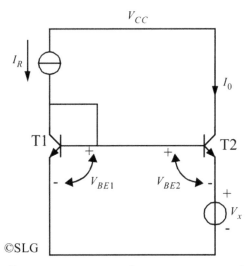

©SLG

Figure 17.13 Trans-linear current generator, where current parameter I_R represents a part of the translation to the **TRH** concentration.

$$V_x = V_{BE1} - V_{BE2} \tag{17.48}$$

$$V_x = V_T \ln\left(\frac{I_R}{I_s}\right) V_T - \ln\left(\frac{I_o}{I_s}\right) \tag{17.49}$$

$$V_x = V_T \left\{ \ln\left(\frac{I_R}{I_s}\right) - \ln\left(\frac{I_o}{I_s}\right) \right\} \tag{17.50}$$

$$V_x = V_T \left\{ \ln\frac{\left(\frac{I_R}{I_s}\right)}{\left(\frac{I_o}{I_s}\right)} \right\} = V_T \ln\left(\frac{I_R}{I_o}\right) \tag{17.51}$$

$$\frac{V_x}{V_T} = \ln\left(\frac{I_o}{I_R}\right) \tag{17.52}$$

$$\frac{I_o}{I_R} = \exp\left(\frac{V_x}{V_T}\right) \tag{17.53}$$

$$I_{C2} = I_0 = I_R \exp\left(-\frac{U_x}{V_T}\right) \tag{17.54}$$

or

$$I_0 = I_R \exp\left(-\frac{V_x}{V_T}\right) \tag{17.55}$$

From Equation (17.55), we appreciate that the output current is an exponential function of V_x.

The input variable U_x/V_T is related to the HP variable φ[FT4] and the output variable I_0 represents the HP variable [TSH]. This trans-linear circuit mimics exactly the HP characteristic. The model parameter φ can be introduced as the parameter coefficient of V_x written as

$$V_x = \varphi U_i V_T, \tag{17.56}$$

which results in

$$I_0 = I_1 \exp\left(-\varphi U_i\right), \tag{17.57}$$

where Equation (17.57) represents the expression for the HP characteristic.

17.6 HPT Functions with Trans-linear Circuits

With a simple modification, we can synthesize the characteristic of the thyroid as follows.

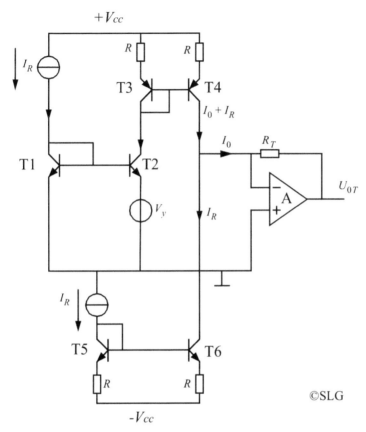

Figure 17.14 Thyroid current expression by subtracting I_R from I_0.

According to the circuit in Figure 17.14, we can write I_0 as

$$I_0 = I_R \exp(-\alpha U_i), \tag{17.58}$$

with

$$V_y = \alpha V_T U_i \tag{17.59}$$

The resulting output current I_T representing the thyroid characteristic is the subtraction of I_R from I_0

$$I_T = I_0 - I_R = I_R \exp(-\alpha U_i) - I_R = -I_R \{1 - \exp(-\alpha U_i)\} \tag{17.60}$$

The output voltage of Figure 17.14 is then

$$U_{0T} = I_R R_T \{1 - \exp(-\alpha U_i)\} \tag{17.61}$$

17.7 The Electronic HP System Implementation

The input and output signals have to be processed with linear analog circuits. These functions are implemented with operational amplifiers. In Figure 17.15, the implementation of the exponential coefficient φ and α can be realized.

When we calculate the equivalent input voltage for the HP section, using Equation (17.55), we now can define the value of the exponential coefficient φ in the equation according to

$$\varphi = \frac{R_2}{R_1 + R_2} \tag{17.62}$$

In our set point example [FT4] = 15 pmol/L and [TSH] = 1.4 mU/L, we have $\varphi = 0.5$ and $S = 2530$.

With the circuit of Figure 17.14, we can write for this set point

$$\frac{V_x}{V_T} = 7.5 \tag{17.63}$$

Choosing R_2 equal to 10 kΩ and $\varphi = 0.4$, we find $R_1 = 15$ kΩ.

For the value of the input voltage, we then find $U_{\text{in}} = 15V_T = 390$ mV.

With $S = 2530$, we can define the equivalent value for I_1 in Figure 17.15 as $I_1 = 2500$ μA.

With the circuit as depicted in Figure 17.16, we can transfer the output current I_0 with a current-to-voltage converter to the appropriate output voltage, representing the value for [TSH], necessary to the input voltage of the thyroid.

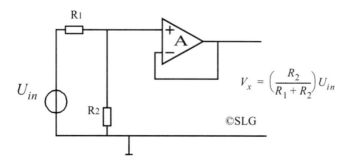

Figure 17.15   Linear input amplifier.

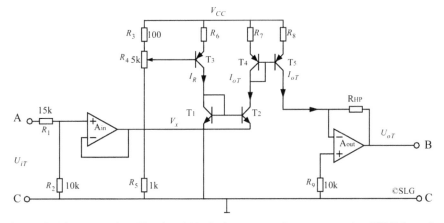

Figure 17.16 HP section. The signal U_{in} is the output voltage representing [FT4] from the thyroid section. The output voltage U_0 represents the control signal [TSH] which is the input for the thyroid section.

The reference current I_R as indicated in Figure 17.16 is generated by T$_3$ where the output current complies to

$$I_{oT} = I_R \exp\left(-\varphi U_{in}\right) \qquad (17.64)$$

The output voltage U_{oT} is then

$$U_{oT} = -I_{oT} R_{HP} = -I_R R_{HP} \exp\left(-\varphi U_{in}\right) \qquad (17.65)$$

The minus sign of this output signal represents the negative feedback factor.

This current is transferred by the current mirror consisting of T$_4$ and T$_5$ where the collector current of T$_5$ is converted by the 100-kΩ resistor of the operational amplifier and provides the signal output $U_0 = 100\,kI_R$. This is the input signal for the thyroid and thus represents the value of [TSH].

With the block function presentation of Figure 17.17, a block-level presentation is introduced. On this abstraction level, we can avoid unnecessary details to give a complete functional block-level schematic in Figure 17.17. From this description we can translate the value of the reference current I_R to the value of [TRH] or model parameter S.

17.8 The Electronic Thyroid Implementation

The thyroid has a similar circuit organization to the HP system of Figure 17.16 and is implemented according to the presentation of

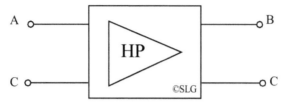

Figure 17.17   Block function presentation of the HP circuit section depicted in Figure 17.12.

Figure 17.17. The input current $I_0 - I_R$ flows into the inverted input of the operational amplifier providing an output voltage U_0.

In Figure 17.18, the exponential function converter is realized by T_1 and T_2 and the current source T_4, the same reference current I_R is distributed by T_3, T_7, and T_8, where the same current is subtracted by T_9. The collector current of T_6 represents the signal output current I_0 which is subtracted by the reference current I_R and added at the inverted input of the operational amplifier A_T. The current input of A_{out} can be written as

$$I_0 - I_R = I_R \exp\left(-\frac{V_y}{V_T}\right) - I_R = -I_R\left\{1 - \exp\left(-\frac{V_y}{V_T}\right)\right\}, \quad (17.66)$$

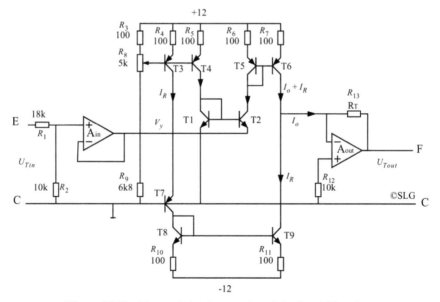

Figure 17.18   Electronic implementation of the thyroid function.

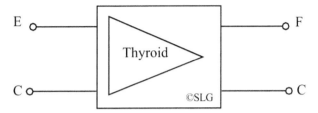

Figure 17.19 Block-level presentation of the thyroid function.

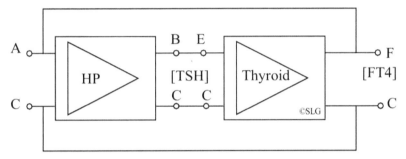

Figure 17.20 Realization of the complete closed-loop HPT hardware simulator.

resulting in

$$U_0 = -R_T I_{in} = R_T I_R \left\{ 1 - \exp\left(-\frac{V_y}{V_T}\right) \right\} \qquad (17.67)$$

This output voltage U_{Tout} will be fed to the input of the HP section with the correct polarity and DC voltage level.

Also for the thyroid, a new level of block schematic is introduced as depicted in Figure 17.19.

The functions from the block-level presentations can be interconnected to represent the complete HPT system as shown in Figure 17.20.

17.9 Discussion

The complete simulator is realized with the mutual connection of the thyroid and HP section. This simulator displays the homeostatic behavior of an HPT system with set point [FT4] = 15 pmol/L and [TSH] = 1.4 mU/L. This results in values for φ = 0.5 and S = 2230 and the thyroid parameters A = 23.7 and α = 0.67.

All mentioned parameters can be modified to other ones resulting in a different point for homeostasis. The electronic realization of the HPT system is actually a hard-wired analog HPT homeostatic point calculator demonstrating the previously theorized principle. Another important property of this electronic functional implementation example is the possibility to model and translate physiological behavior to electronic circuits which opens the all kinds of study opportunities for a certain physiological model without the burden of experiments on human or animal test subjects.

References

[1] Velten, K. (2009). *Mathematical Modeling and Simulation Introduction for Scientists and Engineers*, Weinheim: Wiley-VCH Verlag GmbH & CoKGaA.

[2] Eisenberg, M., Samuels, M., and Extensions. (2008). Validation and clinical applications of a feedback control system simulator of the Hypothalamo-pituitary-thyroid axis. *Thyroid* 18, 1071–1085. doi: 10.1089/thy.2007.0388

[3] Hodgkin, A., Huxley, A., and Katz, B. (1952). Measurement of current – voltage relations in the membrane of loligo. *J. Physiol.* 116, 424–448.

[4] Hodgkin, A. L., Huxley, A. F. (1952). A quantitative description of membrane current and is application to conduction and excitation in nerve. *J. Physiol.* 117, 500–544.

[5] van der Woerd, A. C., and Serdijn, W. A. (1993). Low-Voltage low-power controllable preamplifier for electret microphones. *IEEE. J. Solid State Circuits.* 28:10.

[6] Mulder, J. (1998). *Static and Dynamic Translinear Circuits*. PhD. thesis, University of Technology in Delft, The Netherlands

[7] Aström, K. J., and Murray, R. M. (2009). *FeedBack Systems*, New Jersey: NJ, Princeton University Press.

[8] Keller, F., Emde, C., and Schwarz, A. (1988). Exponential function for calculating saturable enzyme kinetics. *Clin. Chem.* 34, 2486–2489.

[9] Keller, F., and Zellner, D. (1996). The 1-exp function as an alternative model of non-linear saturable kinetics. *Eur. J. Clin. Chem. Clin. Biochem.* 34, 265–271.

[10] Goede, S. L., Leow, M. K., Smit, J. W. A., and Dietrich, J. W. (2014). A novel minimal mathematical model of the hypothalamus–pituitary–thyroid axis validated for individualized clinical applications. *Math. Biosci.* doi: 10.1016/j.mbs.2014.01.001

[11] Goede, S. L., Leow, M. K., Smit, J. W. A., Klein, H. H., and Dietrich, J. W. (2014). Hypothalamus-Pituitary-Thyroid feedback control: implications of mathematical modeling and consequences for thyrotropin (TSH) and free thyroxine (FT4) reference ranges. *Bull. Math. Biol.* doi: 10.1007/s11538-014-9955-5

[12] Leow, M. K., and Goede, S. L. (2014). The homeostatic set point of the hypothalamus-pituitary-thyroid axis – maximum curvature theory for personalized euthyroid targets. *Theor. Biol. Med. Model.* 11:3. doi: 10.1186/1742-4682-11-3

18

From Theory to Practice: Computer-aided Set Point Application

"Experience without theory is blind, but theory without experience is mere intellectual play."

–Immanuel Kant (1724–1804)

18.1 Introduction

Mathematical models remain theoretical constructs as bodies without souls until they are injected with life by being applied and proven to work in the real world. The maximum curvature theory which forms the basis of the euthyroid set point has a sound rationale and logical appeal. It actually predicts the individualized euthyroid set points of thyroid patients with a certain level of precision. This is what we have previously shown which gives us the impetus to develop a software algorithm based on this theory so that it can be easily accessed on laptop and desktop computers or mobile phone platforms by practicing physicians. We envision the day when thyroid patients will routinely be treated to their individualized set points as computed by the software quickly in a busy clinic with a few clicks or taps by their health care providers.

It is unfortunate that the concept of titrating thyroid medications to a personalized set point is as yet largely unknown if not unheard of. Physicians follow present guidelines which continue to promulgate treatment targeted at population ranges of [TSH]. The clinical inspection and interpretation of [FT4] is generally neglected. Until the euthyroid set point is appreciated by the medical fraternity to be unique to every patient, this present approach is sub-optimal. For an engineer it is, however, plain to see that it is an established fact that the individual [FT4] and [TSH] during the healthy state will not change very much over time because this euthyroid set point is probably genetically programmed. That accounts for the very significantly narrow variation in [FT4] and [TSH] for any given person compared with the very wide population normal ranges. The commonly used clinical research standards targeted on group behavioral evaluations with not understood statistical methods can of course not lead to the necessary acceptance. Instead of this method the clinician has to be assisted by an engineering counterpart to provide a proper interpretation of measured results in order to come to a scientifically and clinically balanced view.

However, for the time being, this software may be licensed for use in the clinics only on a case-by-case basis according to each physician's best clinical judgment after having informed the relevant patients that it is still experimental and undergoing research evaluation. Until the set point strategy becomes accepted as standard treatment and integrated into medical practice guidelines, each doctor who treats a thyroid patient using such a method does so with full informed consent from the patient within the context of a doctor–patient relationship and remains bound by the Hippocratic oath to act in the patient's best interest and not to do harm. Of course, medical apps and software may be classified as medical devices and subject to regulatory approval depending on the individual country's laws and regulatory authorities. But where it is permitted, the use of this software must be restricted to patients whose well-being is evidently suboptimal while they are still defined as "euthyroid" according to the population normal ranges.

18.2 Thyroid-SPOT Software

The euthyroid set point algorithm we developed was patented in 2013, after which we proceeded next to produce a computer program that performs this set point computation effortlessly. This software has been codenamed Thyroid-SPOT, an abbreviation for Thyroid Set Point Optimization and Targeting. A desktop version of the Thyroid-SPOT software was developed a few years ago. This uses the algorithm we jointly co-developed which has been patented. The desktop version is presently being licensed for use in some thyroid research projects. More recently, a mobile phone Thyroid-SPOT app has also been developed which facilitates the computation of set points for medical doctors who are on the move from wards to wards in the hospitals or whenever they are being consulted by the phone. The patent and software are based on previous publications of the authors [1–4].

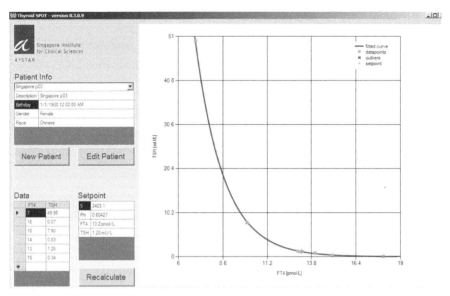

Figure 18.1 Graphical user interface of the desktop version of Thyroid-SPOT.

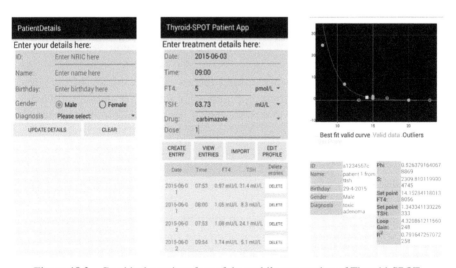

Figure 18.2 Graphical user interface of the mobile app version of Thyroid-SPOT.

18.3 Clinical Studies Using Thyroid-SPOT

With the conduction of a formal randomized double-blind placebo controlled study the effectiveness of this desktop version of Thyroid-SPOT entitled EQUILEBRATE, is evaluated.

The mnemonic which stands for "A Prospective Randomized, Double-blind, Parallel Arm, Multi-center Clinical Trial to **E**valuate the **Qu**ality of **L**ife and **E**uthyroid **B**alance using Conventional Thyroid Hormone **R**eplacement versus Se**t** Point Strategy." The patients recruited for EQUILE-BRATE are mainly those who have primary hypothyroidism being replaced with L-thyroxine.

Although quality of life (QOL) questionaires have a doubtful reputation in serious research methods, the results could possibly be helpful for the completion of the study. Quality of life in this clinical trial is evaluated using a robust questionnaire (SF-36) as well as a thyroid-specific questionnaire called ThyPro.

In yet another study codenamed TRIBUTE which stands for "The Role of **T**hyroid Status in **R**egulating **B**rown Adipose Tiss**u**e Activity. White Adipose **T**issue Partitioning and Resting **E**nergy Expenditure," patients with hyperthyroidism from Graves' disease are being assessed. The assessment is performed for both residual thyroid symptoms and objective measures of fasting lipids, serum electrolytes, body weight, and brown adipose tissue function. These evaluations take place as they undergo anti-thyroid drug therapy.

Finally, the authors are also attempting to prove the accuracy of the set point prediction by comparing the computed set points of thyroidectomised patients on L-thyroxine replacement to the TFTs pre-thyroidectomy at a time when these patients were healthy.

This is done in a retrospective study entitled PREDICT-IT which stands for "**P**rofiling **Re**trospective Thyroid Function **Da**ta **i**n Complete **T**hyroidectomy Patients to **I**nvestigate the hypothalamus–pituitary–thyroid (HPT) axis Set Poin**t**."

The co-author (ML) has so far tried on the iOS version of the app for over a year prior to this manuscript being written. He has found it exceptionally handy when he needs to estimate the dose requirement of some of his thyroid patients. He did this while attending to a phone specialist consultation or whenever he wishes to assess if a given dose of thyroid medication might be either suboptimal or excessive for a given patient located in another hospital or clinic while he attends to medical "on-call" duties based on the TFT results and the set point that has been predicted by the mobile version of Thyroid-SPOT.

18.4 Thyroid-SPOT (Set Point Optimization and Targeting)

The complete theoretical treatment and discussion about the HPT system is too elaborate for direct practical applications. Most of the physicians, clinicians, and patients will not have the background to apply the theoretical fundamentals. On the other hand, the applicability of measured results from TFTs should be handled adequately in order to determine the optimal individual point for [FT4] and [TSH] when they are treated with levothyroxine (L-T4).

The computer tool Thyroid-SPOT can be used for this purpose which helps clinicians, physicians, and patients to find the optimal area for their unique [FT4] and [TSH]. Reference ranges for [FT4] and [TSH] have not to be used anymore, even stronger, should be ignored because they are in this respect obsolete. The individual characteristics as calculated by Thyroid-SPOT will present the optimum solution for the individual in question and can sometimes fall outside the normal reference ranges. The following description will guide the user through the possibilities of this set point calculator.

In endocrinology, we are familiar with the hypothalamus pituitary axis which is a central regulation system for a variety of hormonal controls.

The HPT axis is known as the closed-loop control system for thyroid hormones.

When the thyroid is not active anymore, most individuals can suffer from a hypothyroidism condition. Hypothyroidism is normally treated with the synthetic form of thyroxine also known as L-T4.

During the treatment of hypothyroidism to a more euthyroid condition, TFTs, measuring free T4 (FT4) and thyroid stimulation hormone (TSH) concentrations, are taken with intervals of about 6 weeks.

Recent research has revealed that every person has a unique relationship between the concentrations of measured FT4 and TSH. This relationship is determined by the properties of the hypothalamus–pituitary system.

When we plot the TFT results during L-T4 titration in a plane with a linear horizontal axis for FT4 and a linear vertical axis for TSH, we find a curved characteristic representing all the possible equilibriums for FT4 and TSH,

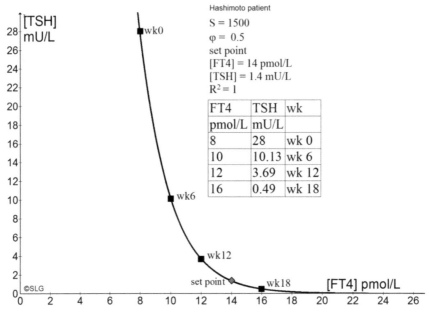

Figure 18.3   A graphical plot of the measured FT4–TSH pairs commonly encountered during the L-T4 titration process of a hypothyroid patient.

regardless the condition of the patient. This means that very low FT4 values (5 pmol/L) can correspond with relatively high values of TSH (50 mU/L) or just as high FT4 values (30 pmol/L) can correspond with very low values of TSH (0.1 mU/L) which are all part of the same curve.

Figure 18.3 depicts such a situation.

Figure 18.3 serves as a representative example of the treatment of a hypothyroid patient and can be used as a template for all other data analyses.

From Figure 18.3, we appreciate that the highest value of TSH is measured in week 0, FT4 = 8 pmol/L and (TSH = 28 mU/L) the moment we start the treatment with L-T4.

After 6 weeks, we find the second pair of FT4 and TSH values, respectively, FT4 = 10 pmol/L and TSH = 10.13 mU/L, after 12 weeks, we have FT4 = 12 pmol/L with a corresponding TSH = 3.69 mU/L, and after 18 weeks, we find FT4 = 16 pmol/L with a corresponding TSH = 0.49 mU/L.

The TFT readings from week 12 and week 18 seem reasonable, because they fall within the standard reference ranges for FT4 and TSH.

However, how can we be sure what the best TFT result will be?

The latest scientific research on this topic revealed that the answer is found on the part of the curve with the strongest bending, or strongest curvature. There is only one point on this characteristic where the curvature is the strongest. This is the point of natural homeostasis for a normal operating euthyroid system. On that particular point, we also find the correct value of FT4 with the belonging TSH value. This is indicated in Figure 18.3 with a black diamond.

The theory resulted in an algorithm that has been incorporated in the clinical support tool Thyroid-SPOT to calculate automatically the set point value.

When the tool is installed on your desktop, you can create and maintain your own database containing the data belonging to your patient.

1. At first, you have to identify the patients name, birthday, gender, type of treatment, etc.
2. Then you save these data.
3. The name of the patient is then visible in your data list.
4. Select this name.
5. Enter the data of the TFT values in the input cadre.
6. When finished, press "**recalculate**."

Then you will find all relevant characteristics of the patient, including the set point.

The set point value is the homeostatic condition where the patient has an optimum setting for the dosage of L-T4 that can realize this value of FT4 and TSH.

In Figure 18.4, you will find an example of the thyroid-SPOT presentation of the example from Figure 18.3.

18.5 Thyroid Function Tests

The quality of TFTs is crucial for a reliable result in Thyroid-SPOT.

Therefore, we recommend the following.

1. Blood sampling for a TFT should preferably be performed between 7.00 and 9.00 h in the morning.
2. The patient should refrain from taking medication and breakfast before blood sampling.
3. After the blood sampling, the patient can take the necessary medication and food.

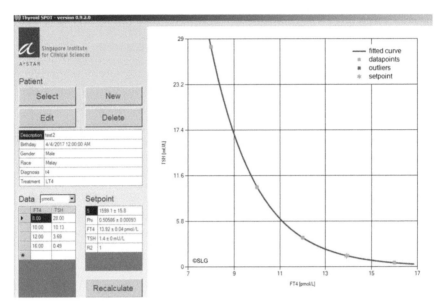

Figure 18.4   A graphical presentation of the thyroid function test (TFT) series inserted and processed with Thyroid-SPOT.

4. All TFT data used in Thyroid-SPOT should be from the same laboratory and the analysis should be performed with the same assays for TSH and FT4.
5. FT4 should be determined with equilibrium dialyses or ultra-filtration to separate the free T4 fraction from the bound fraction.
6. The prepared free fraction should preferably be measured with tandem mass spectrometry or equivalent analysis equipment.
7. The measured value for FT4 has to be presented according to the International Standard of Units in pmol/L with the accuracy of at least one decimal like FT4 = 15.6 pmol/L.
8. Measured values of TSH < 0.1 mU/L will lose their reliability because of the reduced amounts of TSH concentrations and the restricted accuracy of the assay in general. However, small differences in TSH concentrations can be detected like differences of about 0.01 mU/L or even smaller fractions, but the accuracy is a total different issue.
9. Measured values of FT4 concentrations smaller than 1 pmol/L can be considered as unreliable because of the deteriorating accuracy below this value.

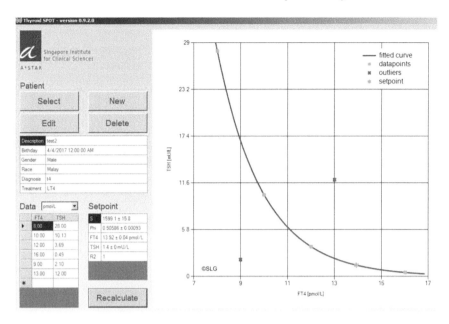

Figure 18.5   A representation of the curve as calculated from the previous TFT values, but now we have added FT4 = 9 with TSH = 2.1 and additionally FT4 = 13 and TSH = 12. These last two TFT values are now indicated as outliers with a red cross and will be neglected in the curve presentation.

Figure 18.5 depicts a representation of the curve as calculated from the previous TFT values with an additional point.

18.6 Interpretation of Test Results

When Thyroid-SPOT provides the calculated result we encounter in general a fitting quality indicated as R^2.

This figure indicates the reliability and accuracy of the calculated curve.

We can accept values like $0.5 < R^2 < 1$ which provides us with an acceptable value of the set point.

Normally, clinicians are educated in the interpretation of TSH concentration values.

However, physiologically the only important variable is FT4 and TSH is only the signaling variable to control the thyroid.

This statement is a paradigm change in the way we have to interpret the HPT system.

Another important qualifier is indicated as φ, representing the steepness of the curve and thus the relatively high variations in TSH when FT4 is varied.

Normally, we can expect values like $0.3 < \varphi < 1$.

Persons with a φ of 0.3 or lower have a relatively low sensitivity for changes in FT4 concentrations because the belonging TSH variations are also relatively low.

These types of curves show a relative high tolerance for deviations from the set point value and individuals with such a curve will generally not suffer from small FT4 set point deviations.

When we encounter curves with φ of 0.5 and higher, we have to deal with an increased sensitivity for all kinds of FT4 variations. Persons with such a high value of φ are extremely sensitive for any L-T4 dose deviation from their set point value!

We have experienced that these persons are responsible for most of the complaints related to an under- or over-dose position of FT4.

As mentioned before, TSH is a control variable and is the sole variable used to determine all functions in the HPT system.

Because of this property, we cannot use TSH as a determining number.

In every individual, the result of TSH means something totally different!

As a striking example, we use here the set point value of TSH ($[TSH]_{sp}$) which determines the value of φ like

$$\varphi = \frac{1}{[TSH]_{sp}\sqrt{2}}.$$

References

[1] Goede, S. L., Leow, M. K., Smit, J. W. A., and Dietrich, J. W. (2014). A novel minimal mathematical model of the hypothalamus–pituitary–thyroid axis validated for individualized clinical applications. *Math. Biosci.* 249, 1–7. doi: 10.1016/j.mbs.2014.01.001

[2] Leow, M. K., and Goede, S. L. (2014). The homeostatic set point of the hypothalamus–pituitary–thyroid axis – maximum curvature theory for personalized euthyroid targets. *Theor. Biol. Med. Model.* 11:35. doi: 10.1186/1742-4682-11-35

[3] Middelhoek, M. G. (1992). *The Identification of Analytical Device Models*. Ph.D. thesis, Delft University Press, Delft.

[4] Leow, M. K., Goede, S. L., and Dietrich, J. W. (2012). System and method for deriving PARAMETERs for homeostasis feedback control of an individual Singapore. Patent Application No 201208940-5.

Appendix

Formula Overview – Thyroid Systems Engineering

The following overview presents all relevant formulae developed in this book.

Hypothalamus–Pituitary (HP) characteristic

$$[TSH] = \frac{S}{\exp(\varphi[FT4])} \tag{A.1}$$

$$\varphi = \left(\frac{S}{([FT4]_1 - [FT4]_2)} \right) \ln \left(\frac{[TSH]_2}{[TSH]_1} \right) \tag{A.2}$$

$$S = [TSH]_1 \exp(\varphi[FT4]_1) = [TSH]_2 \exp(\varphi[FT4]_2) \tag{A.3}$$

$$G_{HP} = \frac{d[TSH]}{d[FT4]} = -\varphi[TSH] \tag{A.4}$$

$$S = \frac{M}{\exp(\varepsilon[FT3])} \tag{A.5}$$

Thyroid (T) characteristic

$$[FT4] = A(1 - \exp(-\alpha[TSH])) \tag{A.6}$$

$$G_T = \frac{d[FT4]}{d[TSH]} = \frac{\alpha A}{\exp(\alpha[TSH])} \tag{A.7}$$

HPT loop gain

$$G_L = \left| \left(\frac{d[FT4]}{d[TSH]} \right)_{HP} \left(\frac{d[FT4]}{d[TSH]} \right)_T \right| = \frac{\varphi \alpha A[TSH]}{\exp(\alpha[TSH])} \tag{A.8}$$

$$GL_{\max} = \frac{\varphi A}{\exp(1)} \tag{A.9}$$

HPT loop stability condition

$$GL_{\max} = \frac{\varphi A}{\exp(1)} = 0.41\frac{[FT4]_{sp}}{[TSH]_{sp}} > 1 \tag{A.10}$$

HP set point relations

$$[FT4]_{sp} = \frac{\ln\left(\varphi S\sqrt{2}\right)}{\varphi} \tag{A.11}$$

$$[TSH]_{sp} = \frac{1}{\varphi\sqrt{2}} \tag{A.12}$$

$$\frac{d[TSH]_{sp}}{d[FT4]_{sp}} = -\frac{1}{\sqrt{2}} = -0.707 \tag{A.13}$$

Thyroid set point relations

$$A = \frac{[FT4]_{sp}}{0.632} \tag{A.14}$$

$$\alpha = \frac{1}{[TSH]_{sp}} \tag{A.15}$$

Thyroid max curvature

$$[TSH] = \frac{\ln\left(\alpha A\sqrt{2}\right)}{\alpha} \tag{A.16}$$

$$[FT4] = \frac{\alpha A\sqrt{2} - 1}{\alpha\sqrt{2}} \tag{A.17}$$

From only one known (measured) value of [FT4] = [FT4]$_{\text{measured}}$.

[FT4] calculation based on chosen reference value of [TSH]$_{\text{ref}}$.

$$[FT4] = [FT4]_{\text{measured}} + [TSH]_{\text{ref}}\sqrt{2}\ln\frac{[TSH]_{\text{measured}}}{[TSH]_{\text{ref}}} \tag{A.18}$$

Pharmacokinetic formulae

Half-life

$$A_t = A_0 \exp(-t/\tau) \tag{A.19}$$

$t_{1/2}$ represents the half-life time in days

$$\tau = \frac{t_{1/2}}{\ln(2)} = 1.4427 t_{1/2} \tag{A.20}$$

Steady-state A_e

$$A_e = \frac{\text{periodical dose in } \mu\text{g over n days}}{1 - \exp(-n/\tau)} \tag{A.21}$$

Continuous time function of the accumulated level

$$\text{Daily dose} = D_d \tag{A.22}$$

$$A_e(t) = D_d + (A_e - D_d)\{1 - \exp(-n/\tau)\} \tag{A.23}$$

$$T4_{\text{average}} = A_e \tau \{1 - \exp(-1/\tau)\} = D_d \tau \tag{A.24}$$

Index

About the Authors

Simon Lucas Goede was born in 1948 in Westzaan, The Netherlands. He studied Chemistry at the Vrije Universiteit Amsterdam for two years and received his MSc in micro-electronics engineering from Delft University of Technology, The Netherlands, in 1979. After mastering the complete scope of advanced mathematics while pursuing his education in nano-electronics engineering, he worked in the micro-electronics and telecommunication industry for 20 years. After his retirement, he studied the endocrine system with respect to the physiology of the thyroid, pancreas, and adrenal glands. He published seven papers related to the hypothalamus–pituitary–thyroid system based on his knowledge of systems theory. The results of these clinically verifiable publications are applicable to the treatment of patients with thyroid disorders. This holds particularly in those with primary hypothyroidism through the elucidation of their individualized homeostatic euthyroid set points which guide optimized medication dosage. Besides thyroid systems research, the author is involved in modeling the glucose/insulin and the adrenal systems where the same modeling principles have been successfully applied.

Simon L. Goede, *MSc*
Systems Research NL
Oterlekerweg 4, 1841 GP Stompetoren, The Netherlands

Melvin Khee Shing Leow is a senior consultant endocrinologist at Tan Tock Seng Hospital, visiting consultant at the Division of Endocrinology at the National University Hospital and the Clinical Trials & Research Unit of Changi General Hospital in Singapore. He is a clinician scientist and is jointly appointed a principal investigator in addition to being the deputy director of the laboratory of the Clinical Nutrition Research Centre located at the Centre for Translational Medicine on the NUS campus. His research focuses on energy metabolism and metabolic physiology including feedback loops regulating the intricate network of the adipose tissue, thyroid, gastrointestinal tract, pancreas, liver, and skeletal muscles. His field of interests includes thyroidology, adipocyte biology, diabetes, endocrinologic oncology, and endocrine manifestations of systemic disorders. In addition, he has a deep interest in mathematics for many years and does mathematical modeling of endocrine physiology. Dr. Leow holds academic and teaching appointments including associate professor in the Lee Kong Chian School of Medicine at Nanyang Technological University, clinical associate professor at Yong Loo Lin School of Medicine, and adjunct associate professor at Duke-NUS Medical School. He is an elected fellow of the American College of Endocrinology, the Academy of Medicine (Singapore), the Royal College of Physicians of Edinburgh, the Royal College of Pathologists of London and Dr. Leow is also the President-elect of the Endocrine and Metabolic Society of Singapore (EMSS).

A/Prof. Melvin Khee Shing Leow
MBBS, MMed (Int Med), MSc, FACE, FAMS, FACP, FRCP (Edin), FRCPath, PhD
Republic of Singapore